U.S. Department
of Transportation

**National Highway
Traffic Safety
Administration**

DOT HS 810 739

NHTSA Technical Report

November 2006

HIC Test Results before and after the
1999-2003 Head Impact Upgrade of FMVSS 201

1. Report No. DOT HS 810 739	2. Government Accession No.	3. Recipient's Catalog No.	
4. Title and Subtitle HIC TEST RESULTS BEFORE AND AFTER THE 1999-2003 HEAD IMPACT UPGRADE OF FMVSS 201		5. Report Date November 2006	
		6. Performing Organization Code	
7. Author(s) Charles J. Kahane, Ph.D. and Marcia J. Tarbet		8. Performing Organization Report No.	
9. Performing Organization Name and Address Evaluation Division; National Center for Statistics and Analysis National Highway Traffic Safety Administration Washington, DC 20590		10. Work Unit No. (TRAIS)	
		11. Contract or Grant No.	
12. Sponsoring Agency Name and Address Department of Transportation National Highway Traffic Safety Administration Washington, DC 20590		13. Type of Report and Period Covered NHTSA Technical Report	
		14. Sponsoring Agency Code	
15. Supplementary Notes			

16. Abstract

Federal Motor Vehicle Safety Standard (FMVSS) 201 – Occupant Protection in Interior Impact – was upgraded in 1995, with a 1998-2002 phase-in, to reduce occupants' risk of head injury from contact during crashes with a vehicle's upper interior, including its pillars, roof headers and side rails, and the upper roof. Initially, energy-absorbing materials alone were used to meet the standard; later, some vehicles were also equipped with head-protection air bags. NHTSA does not yet have enough crash data to evaluate the injury-reducing effectiveness of the energy-absorbing materials. However, the agency has conducted 154 matched pairs of impact tests with free-motion headforms in pre- and post-standard vehicles of 15 selected make-models. The Head Injury Criterion, HIC(d) averaged 909.9 in the 154 pre-standard tests and 667.5 in the post-standard vehicles. This is a statistically significant average improvement of 242.4 units of HIC.

17. Key Words NHTSA; FMVSS; HIC; head injury; padding; A-pillar; statistical analysis; evaluation; effectiveness; crashworthiness;	18. Distribution Statement Document is available to the public at the Docket Management System of the U.S. Department of Transportation, http://dms.dot.gov, Docket Number 27371.		
19. Security Classif. (Of this report) Unclassified	20. Security Classif. (Of this page) Unclassified	21. No. of Pages 47	22. Price

Form DOT F 1700.7 (8-72)   Reproduction of completed page authorized

EXECUTIVE SUMMARY

The purpose of Federal Motor Vehicle Safety Standard (FMVSS) 201 – Occupant Protection in Interior Impact – is to reduce occupants' risk of head injury in crashes. The performance test requirements of FMVSS 201 limit the force allowed when a dummy headform impacts locations in the vehicle's interior that might actually be contacted by occupants' heads during crashes. NHTSA's major upgrade of FMVSS 201 in 1995 added the A-, B- and other pillars; roof headers; roof side rails; and the upper roof to the list of test locations. The new requirements phased in during 1998-2002. Initially, energy-absorbing materials alone were used to meet the standard; later, some vehicles were also equipped with head-protection air bags.

The injury-reducing effectiveness of the energy-absorbing materials will ultimately be evaluated by statistical analyses of crash data. As of November 2006, NHTSA does not yet have nearly enough crash data. In the interim, this study documents the energy-absorbing materials actually installed in production vehicles and evaluates their effect on performance in laboratory tests. The materials – located beneath the potential impact locations – include composite plastic foam padding, injection-molded ribs or ridges in parallel or egg-crate-like configurations, crushable tubes and more flexible designs for interior surfaces and components.

NHTSA purposively selected 15 high-sales make-models of passenger cars, pickup trucks, SUVs and minivans that had been tested for compliance with FMVSS 201. A contractor purchased 15 pre-standard vehicles of the same make-models, and performed identical headform impact tests, at the same locations, as in the compliance tests. (When an exact match was impossible because a make-model was discontinued or its interior redesigned, the closest corresponding model or location was selected.) In all, there were 154 matched pairs of impact tests in pre- and post-standard vehicles.

The Head Injury Criterion, HIC(d) measured on the dummy headforms has been validated as a criterion for predicting head injury risk in crashes. For compliance with FMVSS 201, HIC(d) must be 1000 or less at each tested location.

- HIC(d) averaged 909.9 in the 154 individual pre-standard impact tests and 667.5 in the post-standard tests. This is a statistically significant average improvement of 242.4 units of HIC(d).

- HIC(d) exceeded 1000 in 47 of the 154 locations tested in pre-standard vehicles, but was less than 1000 in each of the 154 locations in the post-standard vehicles.

- Only 2 of the 15 pre-standard vehicles, but all of the post-standard vehicles had HIC(d) ≤ 1000 at each location.

The analyses demonstrate a substantial improvement of HIC(d) after the FMVSS 201 upgrade for 15 high-sales make-models. They provide the most positive evidence available, for the time being, that the energy-absorbing materials introduced in response to FMVSS 201 are likely reducing head injuries in crashes.

HIC TEST RESULTS BEFORE AND AFTER THE
1999-2003 HEAD IMPACT UPGRADE OF FMVSS 201

The 1999-2003 head impact upgrade of FMVSS 201

Federal Motor Vehicle Safety Standard (FMVSS) 201 – Occupant Protection in Interior Impact – "specifies requirements to afford impact protection for occupants."[1] Over the years, FMVSS 201 has primarily consisted of performance requirements limiting the amount of resistive force allowed when a headform is impacted into various sections of the vehicle interior that are typically contacted by occupants' heads during crashes.

FMVSS 201 was one of NHTSA's initial safety standards, effective for passenger cars on January 1, 1968 and LTVs ("light trucks and vans," including pickup trucks, vans and SUVs) up to 10,000 pounds Gross Vehicle Weight Rating (GVWR) on September 1, 1981. The standard originally incorporated the Society of Automotive Engineers' (SAE) 15 mph headform impact test, applying it to components identified, from the limited data available at that time, as likely head-contact areas: the top of the instrument panel, seat backs, sun visors, armrests and other projections in head impact areas. The final rule allowed a peak deceleration of the headform of 80 g's over 3 milliseconds on the impact tests. Most cars were apparently meeting the various head-impact requirements of FMVSS 201 well before 1968.[2]

Improved crash data such as the National Automotive Sampling System (NASS) clarified that head impacts with the upper interior of vehicles continued to be the leading cause of fatal head injury for non-ejected occupants, despite the original FMVSS 201. Moreover, the injuries involved components not covered by FMVSS 201, such as pillars, roof side rails and headers, and the roof itself – and many of the injuries would not be mitigated by frontal air bags or increased use of safety belts. Based on 1988-1993 NASS data, NHTSA's 1995 *Final Economic Assessment* estimated that these impacts resulted in 2,170 fatalities and 3,630 serious non-fatal injuries per year to occupants of passenger cars and LTVs.[3]

On August 14, 1995, NHTSA issued a Final Rule extending the head injury protection requirements of FMVSS 201 to new target areas. The existing requirements of FMVSS 201 remain for the original target areas. However, the new target areas in the vehicle's upper interior include the A-, B- and other pillars, the front and rear roof header, the roof side rails, and the upper roof, among others. The speed for the free-motion headform (FMH) impact test for the new areas is 24 km/h (15 mph, as in the original FMVSS 201) but for these targets, the Head Injury Criterion (HIC) may not exceed 1000 for any 36-millisecond period. Impacts may be directed from a range of vertical and horizontal angles. Manufacturers were offered a choice of several alternative phase-in schedules from September 1, 1998 to September 1, 2002. For

---

[1] *Code of Federal Regulations*, Title 49, Government Printing Office, Washington, 2006, Part 571.201.
[2] Kahane, C.J., *An Evaluation of Occupant Protection in Interior Impact for Unrestrained Front Seat Occupants of Cars and Light Trucks*, NHTSA Technical Report No. DOT HS 807 203, Washington, 1988, pp. 2-3; Campbell, B.J., *A Study of Injuries Related to Padding on Instrument Panels*, HSL Publication No. 00427812, Report No. VJ-1823-R2, Cornell Aeronautical Laboratory, Buffalo, 1963; *Federal Register* 31 (December 3, 1966): 15212; *1967 SAE Handbook*, Society of Automotive Engineers, New York, 1967, pp. 881-884.
[3] *Final Economic Assessment, FMVSS No. 201, Upper Interior Head Protection*, NHTSA Plans and Policy, Washington, 1995, p. I-1.

example, they could certify the new requirements on at least 10 percent of cars and LTVs manufactured from September 1, 1998 through August 31, 1999; at least 25 percent of cars and LTVs manufactured from September 1, 1999 through August 31, 2000; at least 40 percent of cars and LTVs manufactured from September 1, 2000 through August 31, 2001; at least 70 percent of cars and LTVs manufactured from September 1, 2001 through August 31, 2002; and all cars and LTVs manufactured on or after September 1, 2002.[4]

In summary, the principal differences between the 1995 and 1968 versions of FMVSS 201 are:

- Additional target areas: pillars, roof side rails and headers, and the roof itself.
- For these additional target areas,
    - HIC may not exceed 1000 for any 36-millisecond period (as opposed to a limit of 80 g's for 3 milliseconds on the original target areas).
    - Impacts may be directed from a range of vertical and, in some cases, horizontal angles.

The 1995 upgrade of FMVSS 201 is a major regulation, and its evaluation is a high priority for NHTSA.[5] Fundamentally, head impact protection could be improved by the addition of energy-absorbing materials, the installation of head-protection air bags, or a combination of both. Head-protection air bags include head curtains, inflatable tubular structures and torso/head combination bags. NHTSA plans to issue statistical analyses of crash data from the Fatality Analysis Reporting System (FARS) and the General Estimates System (GES) that estimate the fatality-reducing effectiveness of head-protection air bags for front-seat occupants in side impacts.[6]

Head-protection air bags, however, were not yet available when FMVSS 201 was upgraded in 1995. They were first offered on some cars in 1998. Even in model year 2003, the first full year after the phase-in of the FMVSS 201 upgrade, over 80 percent of new cars and LTVs were not equipped with head-protection air bags and relied exclusively on energy-absorbing materials to meet the standard; as of model year 2006, the majority of new vehicles still did not have head-protection air bags. The fatality- or injury-reducing effectiveness of these materials cannot be readily estimated from FARS and will require more detailed data that will need some years to accumulate. For the time being, we do not need crash data to address two evaluation questions about the vehicles that meet the FMVSS 201 upgrade **without head-protection air bags**:

---

[4] *Federal Register* 60 (August 18, 1995): 43031. On July 29, 1998, NHTSA amended FMVSS 201 to facilitate the introduction of head-protection air bags – *Federal Register* 63 (August 4, 1998): 41451. Recognizing that the 24 km/h (15 mph) headform test might be a problem in target areas where the undeployed air bag is stored (and, furthermore, an inappropriate test if the bag usually deploys at that speed), NHTSA offered an alternative compliance procedure. Manufacturers have the option to reduce the speed of the headform test to 19.2 km/h (12 mph) on target areas where the bag is stored, provided they can meet a 28.8 km/h (18 mph) lateral (90 degree) crash test for the full vehicle into a pole – with HIC < 1000. The pole test simulates a head impact with the deployed bag.
[5] *National Highway Traffic Safety Administration Evaluation Program Plan, Calendar Years 2004-2007*, NHTSA Report No. DOT HS 809 699, Washington, 2004, p. 8.
[6] Kahane, C.J., *An Evaluation of Side Impact Protection – FMVSS 214 TTI(d) Improvements and Side Air Bags*, NHTSA Technical Report, to appear in 2007.

- In what ways were vehicle components modified?
- Did HIC performance on the 24 km/h impact test improve?

In what ways were vehicle components modified?

In 2003, a NHTSA contractor performed a cost teardown study of the manufacturing and consumer costs of the changes made by the automotive industry to meet the FMVSS 201 upgrade without head-protection air bags.[7] In addition to providing cost estimates, the study describes in detail what energy-absorbing materials were actually added or modified. The contractor studied pre-standard passenger vehicles of ten make-models and post-standard vehicles of the same or corresponding make-models. The vehicles comprise a variety of manufacturers and included six passenger cars, a pickup truck, an SUV and two minivans. In addition to the contractor's own examinations of the components, the contractor received detailed information from the manufacturers, such as parts lists, identifying how vehicles were modified.

VEHICLES INCLUDED IN THE COST ANALYSIS				
Make	**Model**	**Vehicle Type**	**Pre-Standard**	**Post-Standard**
Dodge	Caravan	Minivan	2000	2001
Ford	Crown Victoria	Car	2000	2001
Ford	F-150	Pickup	1999	2000
Ford	Taurus	Car	1998	2001
Honda	Accord	Car	1998	2001
Jeep	Grand Cherokee	SUV	1998	1999
Kia	Spectra	Car	2000	2002
Pontiac	Montana	Minivan	1998	2001
Toyota	Camry	Car	1998	2002
VW	Jetta	Car	1998	2002

Approaches used to meet the standard include composite plastic foam padding, injection-molded ribs in parallel or egg-crate-like configurations (see Figures 1 and 2), ridges molded from composite plastic materials (Figure 3), crushable O-Flex tubes (Figure 4), stretchable fabric materials, and configuration changes of the outer trim parts to allow for the flexing of the part under load to absorb part of the impact energy.

The most popular approaches used by the manufacturers were foam padding and internal collapsible ribs. Other than configuration changes, the least expensive approach was foam padding, which is cut from pre-formed rolls of material and glued in place.

---

[7] Ludtke, N.F., Osen, W., Gladstone, R., and Lieberman, W., *Perform Cost and Weight Analysis, Non Air Bag Head Protection Systems, FMVSS 201*, NHTSA Technical Report No. DOT HS 809 810, Washington, 2003.

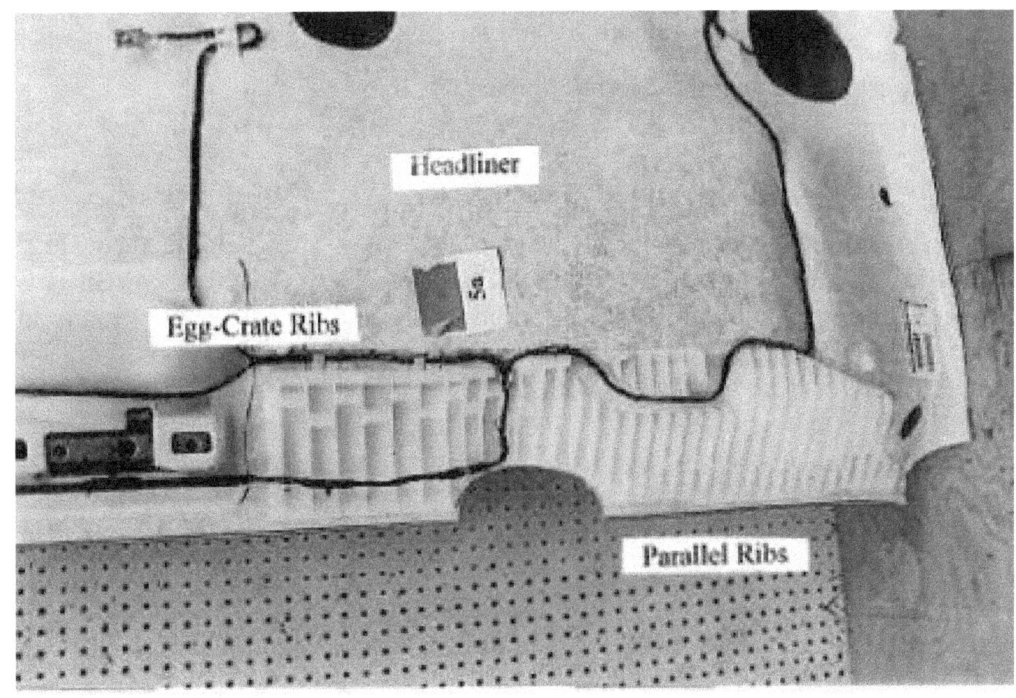

Figure 1: Egg-Crate and Parallel Ribs Built into the Headliner[8]

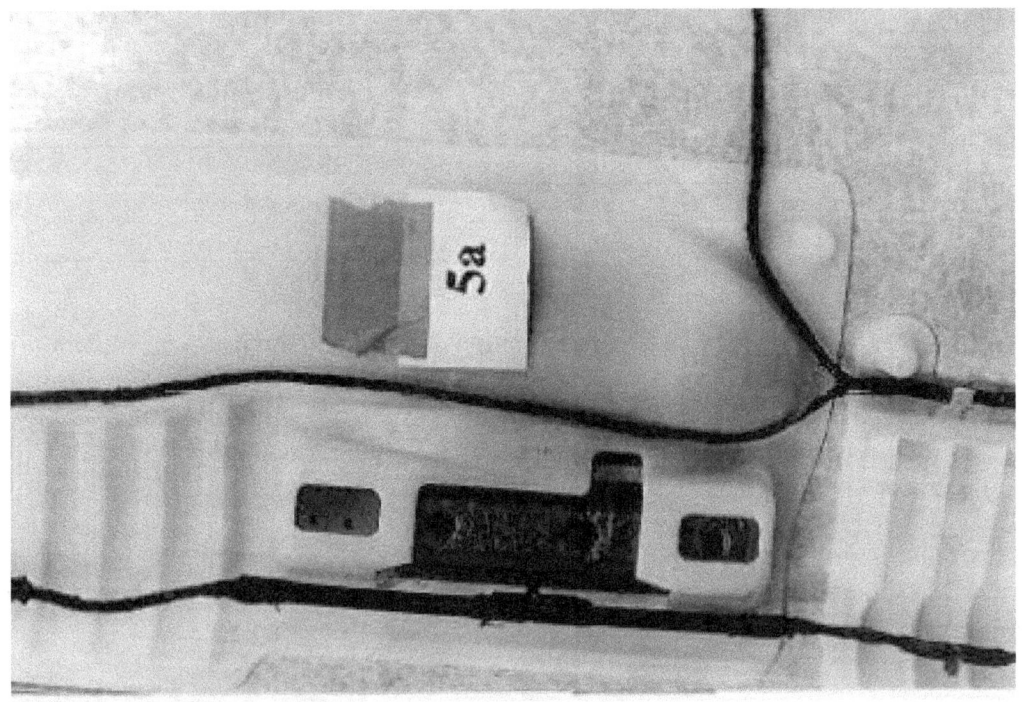

Figure 2: Close-Up of Parallel and Egg-Crate Ribs[8]

---

[8] *Ibid.*, p. 2-6

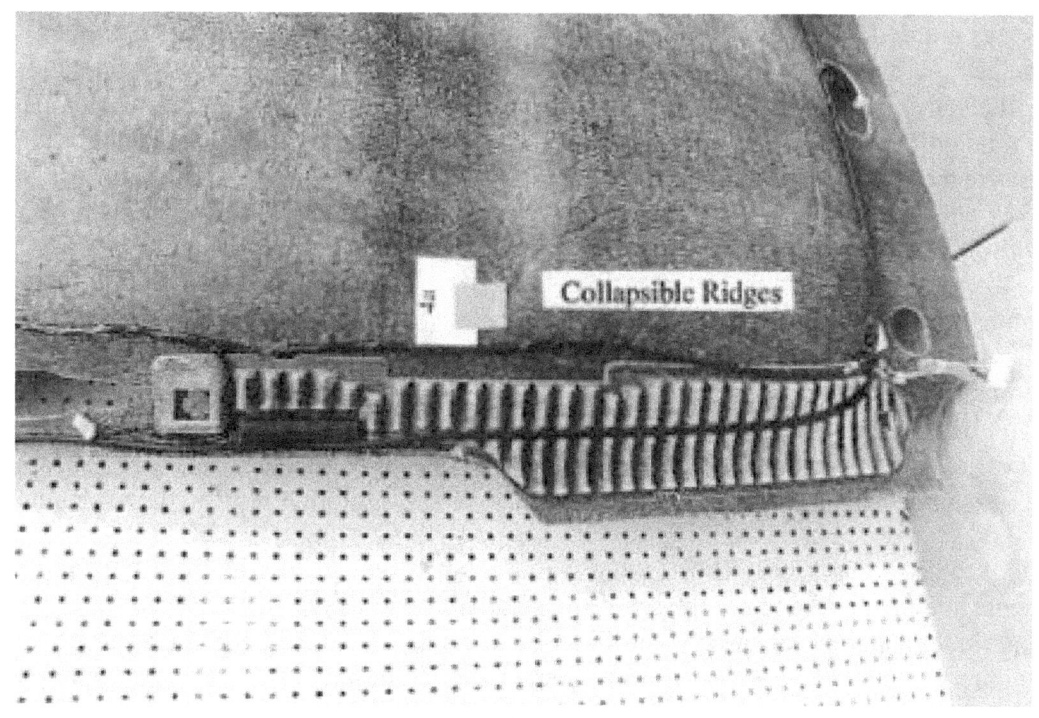

Figure 3: Collapsible Ridges[9]

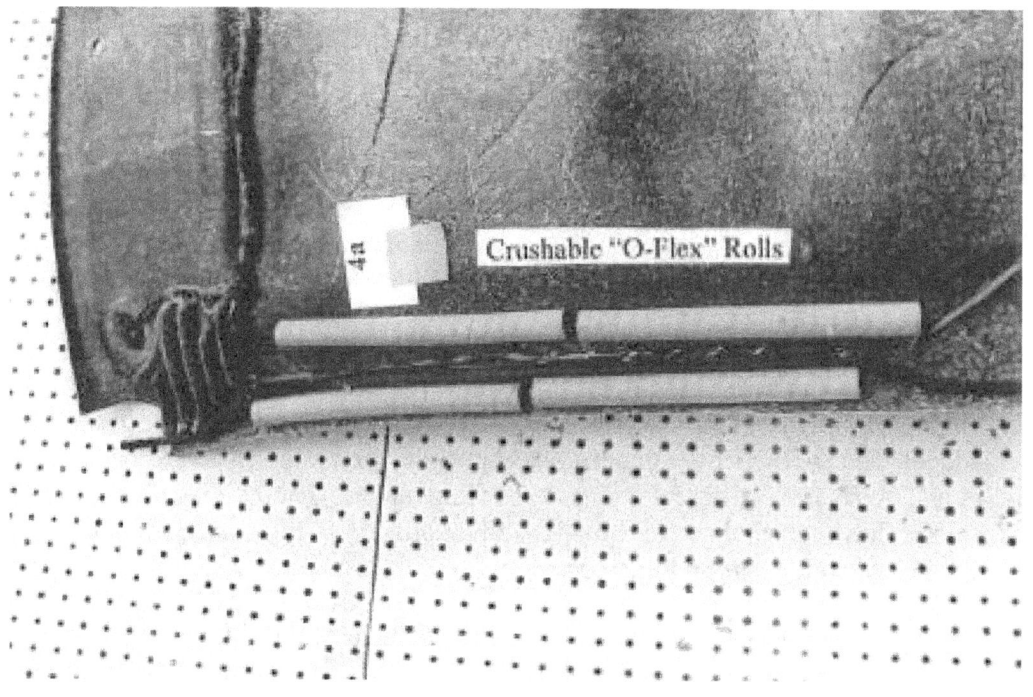

Figure 4: Crushable O-Flex Rolls[9]

[9] *Ibid.*, p. 2-7.

Ridges were molded into some of the foam padding. A cross-section of these ridges presents a pyramid shape with a rounded top. The energy absorption curve generated by foam padding is a straight line starting at zero (first indication of impact load) to a level of the impact load at the point of total collapse. This approach is approximately 50 percent as efficient as the square wave generated by the rib configuration.

The injection-molded ribs in the egg-crate or parallel configuration are the most expensive; however, they are also the most efficient use of material. The foam part returns to its original configuration after the impact. The ribs are thin-walled panels with parallel sides molded with the wall configuration in line with the direction of the expected impact load. The energy absorption curve generated by the collapse of the rib under impact load approaches a square wave (which is the most efficient energy absorption method). The egg-crate rib configuration is the most efficient load absorption design because the ribs are connected at 90° angles, reinforcing the load resistance capability. Parallel ribs do not have the egg-crate reinforcement feature.

Analysis of the cost teardown indicates that manufacturers redesigned their pillar trim components and headliners to comply with the standard on seven of the ten make-models. The remaining three make-models used a combination of redesign, added padding, and ribs for their post-standard vehicles. There were no changes to the C-Pillar and D-Rings except in the Ford vehicles. All internal ribs are made of a collapsible plastic composite material in an egg-crate (honeycomb) and/or parallel configuration. The following paragraphs describe the different approaches used for the ten vehicles evaluated during this project.

Dodge Caravan. The A-Pillar and trim components, B-Pillar and trim components, and headliner have been redesigned to comply with the standard. The cross-section of the A-Pillar has been made deeper and ribs, foam strips, and a ribbed insert have been added. The B-Pillar has a slightly different shape (including a larger flare where the pillar meets the roof), a loop to retain the seat belt, and two added foam pads inside the lower end of the pillar. The headliner – the interior lining of the roof – has six additional plastic panels (combination parallel and egg-crate configuration) across the front and down both sides. Three of these panels serve as both air ventilation ducts and energy absorption devices. The other three panels are devoted to energy absorption purposes. Two small Styrofoam pads were added to the headliner near the top of the B-Pillar.

Ford Crown Victoria. The upper interior components have been redesigned to comply with the standard. The A-, B-, and C-Pillar trim has been changed and internal parallel ribs and fasteners have been added. The D-Ring cover material has been changed to include internal ribs. The grab handles and hooks have different energy absorbing materials and collapsible fasteners. The headliner material has been changed and foam padding added. O-Flex crushable tubes and egg-crate pieces have been added around the side edges.

Ford F-150 pickup truck. The upper interior components have been redesigned to comply with the standard. The A- and B- Pillar trim has been changed and internal parallel ribs and fasteners have been added. In addition, an O-Flex crushable tube has been added to the B-Pillar trim. The D-Ring covers are made of a different material that includes internal collapsible ribs. The grab handle material has been changed to an energy absorbing type and is supported by a bracket attached to the inner A-Pillar structure. Blocks of foam padding and O-Flex crushable tubes have been added to selected areas around the outer edge of the headliner, and a rib cartridge has been added to the foam at the side rail near the A-Pillar. The overhead console is made of a new material that has better performance at impact.

Ford Taurus. The upper interior components have been redesigned to comply with the standard. The shape of the A-Pillar has been changed. Foam padding inside the trim and collapsible fasteners has been added to the A-Pillar trim. A metal strap and washer protect the foam padding from a sheet metal flange. The B-Pillar trim material and shape has been changed. Internal parallel ribs and fasteners have been added to the C-Pillar. The D-Ring covers are made of a different material that includes internal collapsible ribs. The headliner is made of a different material that is slightly thicker than that of the pre-standard model, and three blocks of foam have been placed adjacent to the side rails. These blocks are about six inches in width and the combined length extends from the front to the rear of the headliner.

Honda Accord. The headliner has been redesigned and padding has been added to the A-Pillar trim to comply with the standard. The padding is made from a composite plastic material. Collapsible vertical plastic ridges have been added to the outer edges on the rear of the headliner, while plastic O-Flex crushable tubes have been added to the outer edges on the front. There has been no change to the B-Pillar.

Jeep Grand Cherokee. The A-Pillar, B-Pillar, and headliner have been redesigned to comply with the standard. The cross-section of the A-Pillar has been widened (3½ inches vs. 2 inches). An extra 12 internal egg-crate ribs, with a strip of foam along the inside edge of the pillar, have been added. The B-Pillar also has a wider cross-section (about ½ inch) and an additional 34 internal parallel ribs. The post-standard headliner has nine added ribbed panels, made of molded plastic collapsible foam with ridged features, which are glued to the roof side of the panel.

Kia Spectra. Padding has been added to the A-Pillar trim, B-Pillar trim, and headliner to comply with the standard. A Styrofoam pad is glued inside the A- and B-Pillar trim. Five foam pads have been added to the left and right side of the headliner and glued to the inside. The pads are not a molded shape but simply rectangular blocks sheared from a one-inch thick sheet of foam.

Pontiac Montana. The A-Pillar, B-Pillar, and headliner have been redesigned to comply with the standard. The A-Pillar trim panel cross-section shape is ½ inch wider with small thickness changes. There are 26 additional internal parallel ribs, and a triangular foam pad has been added near the base of the panel. The B-Pillar has an additional 17 internal parallel ribs, a larger cross-section or "footprint" where the B-Pillar meets the roof, and an enlarged upper attaching point that has an additional seven internal parallel ribs. The headliner has an additional six plastic risers, which look like heavy duty "bubble wrap" and hold the foam pads in place. The center seat head protection pads are Styrofoam that are glued directly to the headliner

Toyota Camry. The upper interior components have been redesigned to comply with the standard. The A-Pillar trim has been changed to include internal egg-crate ribs, while the B-Pillar trim has been changed to include internal parallel ribs and reinforcement padding. Changes have been made in the shapes of the rear header, the roof side rails, and the roof upper header. Plus, reinforcements have been added to the roof side rails to improve their head protection capability.

<u>Volkswagen Jetta.</u> The upper interior components have been redesigned to comply with the standard. The post-standard A-Pillar extends from the dash to the roof, whereas the pre-standard pillar had extended from the floor to the roof. The cross-section of the A-Pillar trim panel has been enlarged and is covered with a stretchable fabric over a 1/32-inch thick felt. The shape of the cross-section of the B-Pillar has also been changed and two Styrofoam strips have been added. The trim panel is covered with the same stretchable fabric as the A-Pillar. The headliner has been redesigned to make it wide enough to reach around the roof outer perimeter structure to afford better head protection. The increase in material for the headliner is nearly 16%.

The next six tables rearrange the preceding information. They examine one location in the vehicle at a time – e.g., the A-Pillar – and describe how the design and materials have changed across the ten make-models in the sample. Fundamentally,

- The A-Pillar was substantially modified in every vehicle. Generally, though, it is not the steel structures of pillars that were modified, but rather their interior lining or "trim panels" – the material between the pillars and the occupant compartment.

- The B-Pillar and headliner (interior lining of the roof) were modified in almost every vehicle.

- Other pillars, roof side rails, the front header and the rear header remained largely unchanged.

MY	Make	Model	Type	A-Pillar Changes
2001	Dodge	Caravan	Van	Redesign. Added ribs, ribbed insert, and foam strips.
2001	Ford	Crown Victoria	PC	Redesign. Added internal collapsible ribs and fasteners.
2000	Ford	F-150	LT	Redesign. Added internal collapsible ribs and fasteners.
2001	Ford	Taurus	PC	Redesign. Changed shape. Added foam padding.
2001	Honda	Accord	PC	Added padding.
1999	Jeep	Grand Cherokee	SUV	Redesign. Changed cross section width. Added ribs and foam.
2002	Kia	Spectra	PC	Added Styrofoam padding.
2001	Pontiac	Montana	Van	Redesign. Widened cross-section shape. Added ribs and triangular foam pad.
2002	Toyota	Camry	PC	Redesign. Added internal collapsible ribs.
2002	VW	Jetta	PC	Redesign. Changed length of trim panel. Enlarged cross section and covered it with stretchable polyester-like fabric over an under layer of felt.

MY	Make	Model	Type	B-Pillar Changes
2001	Dodge	Caravan	Van	Redesign. Changed shape. Added foam pads.
2001	Ford	Crown Victoria	PC	Redesign. Added internal collapsible ribs and fasteners.
2000	Ford	F-150	LT	Redesign. Added internal collapsible ribs, fasteners, and O-Flex tube.
2001	Ford	Taurus	PC	Redesign. Changed shape. Added internal collapsible ribs.
2001	Honda	Accord	PC	Unchanged.
1999	Jeep	Grand Cherokee	SUV	Redesign. Widened cross section. Added internal collapsible ribs.
2002	Kia	Spectra	PC	Added Styrofoam padding.
2001	Pontiac	Montana	Van	Redesign. Enlarged cross-section. Added plastic risers and foam.
2002	Toyota	Camry	PC	Redesign. Added internal collapsible ribs and reinforcement padding.
2002	VW	Jetta	PC	Redesign. Changed shape. Added Styrofoam strips. Covered panel with stretchable polyester-like fabric over an under layer of felt.

MY	Make	Model	Type	Other Pillar Changes
2001	Dodge	Caravan	Van	Unchanged.
2001	Ford	Crown Victoria	PC	Redesign. Added internal collapsible ribs and fasteners.
2000	Ford	F-150	LT	Unchanged.
2001	Ford	Taurus	PC	Redesign. Added internal collapsible ribs and fasteners.
2001	Honda	Accord	PC	Unchanged.
1999	Jeep	Grand Cherokee	SUV	Unchanged.
2002	Kia	Spectra	PC	Unchanged.
2001	Pontiac	Montana	Van	Unchanged.
2002	Toyota	Camry	PC	Unchanged.
2002	VW	Jetta	PC	Unchanged.

MY	Make	Model	Type	Upper Roof (Headliner) Changes
2001	Dodge	Caravan	Van	Redesign. Added plastic panels and Styrofoam pads.
2001	Ford	Crown Victoria	PC	Redesign. Changed headliner material. Added O-Flex crushable tubes, egg-crate (honeycomb) pieces, and foam padding.
2000	Ford	F-150	LT	Redesign. Changed headliner material. Added O-Flex crushable tubes and foam padding.
2001	Ford	Taurus	PC	Changed headliner material. Added foam padding.
2001	Honda	Accord	PC	Redesign. Added plastic ridges and O-Flex crushable tubes.
1999	Jeep	Grand Cherokee	SUV	Redesign. Added plastic ribbed panels.
2002	Kia	Spectra	PC	Added foam pads.
2001	Pontiac	Montana	Van	Redesign. Added 6 plastic risers.
2002	Toyota	Camry	PC	Redesign. Changed shape of roof upper header.
2002	VW	Jetta	PC	Redesign. Widened headliner. Material usage increase was nearly 16%.

MY	Make	Model	Type	Roof Side Rail Changes
2001	Dodge	Caravan	Van	Unchanged.
2001	Ford	Crown Victoria	PC	Unchanged.
2000	Ford	F-150	LT	Unchanged.
2001	Ford	Taurus	PC	Unchanged.
2001	Honda	Accord	PC	Unchanged.
1999	Jeep	Grand Cherokee	SUV	Unchanged.
2002	Kia	Spectra	PC	Unchanged.
2001	Pontiac	Montana	Van	Unchanged.
2002	Toyota	Camry	PC	Redesign. Changed shape. Added reinforcements.
2002	VW	Jetta	PC	Unchanged.

MY	Make	Model	Type	Front and Rear Header Changes
2001	Dodge	Caravan	Van	Unchanged.
2001	Ford	Crown Victoria	PC	Unchanged.
2000	Ford	F-150	LT	Unchanged.
2001	Ford	Taurus	PC	Unchanged.
2001	Honda	Accord	PC	Unchanged.
1999	Jeep	Grand Cherokee	SUV	Unchanged.
2002	Kia	Spectra	PC	Unchanged.
2001	Pontiac	Montana	Van	Unchanged.
2002	Toyota	Camry	PC	Redesign. Changed shape of front and rear headers.
2002	VW	Jetta	PC	Unchanged.

Did HIC performance on the 24 km/h impact test improve?

FMVSS 201 compliance tests The starting point for the analysis are the compliance test results for cars and LTVs certified to meet the upgraded FMVSS 201. NHTSA tested 68 vehicles during Fiscal Years (FY) 1999-2003.[10] The test procedure, TP201U-01 of April 3, 1998, defines up to 24 target locations for headform impacts on the left side of a vehicle and 24 corresponding locations on the right side.[11] Moreover, many of the targets allow choices from a range of horizontal and/or vertical impact angles.[12] The compliance tests, however, do not involve all the available locations, but a selection of approximately ten impacts per vehicle. Target selections vary from vehicle to vehicle, but over the course of compliance testing since 1999 have regularly covered the available locations. Targets for an individual vehicle typically include at least two of the available locations each on the A-Pillar and B-Pillar, and at least one location each on another pillar, the side rail and the roof.

---

[10] Test results accessible on the NHTSA web site at
http://www.nhtsa.dot.gov/cars/testing/comply/S201U/index html.
[11] *Laboratory Test Procedure for FMVSS 201U – Occupant Protection in Interior Impact – Upper Interior Head Impact Protection*, Procedure No. TP201U-01, NHTSA Office of Vehicle Safety Compliance, Washington, April 3,1998, p. 37, accessible from http://nhtsa.gov/portal/site/nhtsa/menuitem.b166d5602714f9a73baf3210dba046a0/.
[12] *Ibid.*, p. 30.

The key parameter is the Head Injury Criterion measured during the headform impact: HIC(d). Four parameters describe the impact: the selected target location (e.g., "AP1" is location no. 1 on the A-Pillar, near the top), the horizontal impact angle, the vertical impact angle, and the velocity. The nominal test velocity is 24 km/h (15 mph), but the actual velocities in compliance tests may vary slightly, usually on the low side.[13] If HIC(d) at every tested location is 1000 or less, the vehicle has passed the compliance test.

Make-models included in the analysis Fifteen make-models were purposively selected for the analysis from the 68 vehicles tested in FY 1999-2003. Although not a probability sample of make-models, they comprise a variety of manufacturers and they include passenger cars, pickup trucks, SUVs and minivans. They are high-sales models that encompassed 23 percent of the new cars and LTVs sold in the United States in model year 2002[14]:

1999	Jeep Grand Cherokee	SUV
2000	Dodge Neon 4-door	passenger car
2002	Dodge Grand Caravan	minivan
2002	Ford F-150 supercab	pickup truck
2002	Ford Explorer 4-door	SUV
1999	Ford Windstar	minivan
2001	Buick LeSabre 4-door	passenger car
1999	Chevrolet Silverado x-cab	pickup truck
2002	Chevrolet Trailblazer 4-door	SUV
2001	Nissan Sentra 4-door	passenger car
1999 & 2003	Honda Accord 4-door	passenger car
2003	Toyota Corolla 4-door	passenger car
1999 & 2002	Toyota Camry 4-door	passenger car
2003	Toyota Tacoma xtracab	pickup truck
2002	Kia Spectra 4-door	passenger car

Only the 2003 Honda Accord was equipped with head-protection air bags. Some of the tests on this vehicle were conducted at 19.2 km/h (12 mph) because the air bags were beneath the target location; these tests were excluded from the analysis. On the Accord and Camry, if the same location was tested on the earlier and later vehicle, the average of the two HIC scores was used in the analysis.

Corresponding pre-standard vehicles In 2004, a NHTSA contractor located and purchased 15 used pre-standard vehicles of the same or closely corresponding make-models at dealerships in the contractor's metropolitan area. The upper interior of the specimen vehicles had to consist completely of its original, undamaged components. The vehicle had to be in good driving condition with body undamaged except for, possibly, minor damages in the front or rear.

---

[13] As discussed above, a 19.2 km/h test can be used if parts of a head-protection air bag are underneath the target location.

[14] All of the passenger cars are 4-door models. Two-door cars were not intentionally excluded, but were not required to be included, either (unlike pickup trucks, SUVs and minivans). None was selected because the overwhelming majority of compliance-tested cars, and almost all the high-sales make-models were 4-door.

Appendix A lists NHTSA's specifications for the used vehicles. The pre-standard vehicles were of model years 1996-1998 – i.e., 6 to 8 years old.[15]

If the make-model had not been restyled (e.g., changed in wheelbase) for two or more years before FMVSS 201 certification, the pre-standard vehicle had to be of the same "generation." If the make-model had been restyled upon FMVSS 201 certification or within a year of it, the pre-standard vehicle had to be of the immediately preceding generation. Here are the 15 post-standard make-models in the analysis and their corresponding pre-standard vehicles:

	Post-Standard Vehicle			Pre-Standard Vehicle	
1999	Jeep Grand Cherokee		1996	same make-model	
2000	Dodge Neon 4-door		1997	same make-model	
2002	Dodge Grand Caravan		1996	same make-model	
2002	Ford F-150 supercab		1998	same make-model	
2002	Ford Explorer 4-door		1998	same make-model	
1999	Ford Windstar		1998	same make-model	
2001	Buick LeSabre 4-door		1998	same make-model	
1999	Chevrolet Silverado x-cab		1998	Chevrolet C-1500 x-cab	
2002	Chevrolet Trailblazer 4-door		1996	Chevrolet Blazer 4-door	
2001	Nissan Sentra 4-door		1998	same make-model	
1999 & 2003	Honda Accord 4-door		1996	same make-model	
2003	Toyota Corolla 4-door		1998	same make-model	
1999 & 2002	Toyota Camry 4-door		1998	same make-model	
2003	Toyota Tacoma xtracab		1997	same make-model	
2002	Kia Spectra 4-door		1997	Kia Sephia 4-door	

Chevrolet Silverado, Chevrolet Trailblazer and Kia Spectra met FMVSS 201 from their first year of production. The Chevrolet C-1500, Chevrolet Blazer and Kia Sephia are the immediate "predecessors" of these vehicles. Otherwise, the pre-standard vehicles were the same make-model as the 201-certified vehicle.

The contractor performed head impact tests on the 15 pre-standard vehicles, following the same test procedure TP201U-01 that had been applied in the compliance tests on the post-standard vehicles. Moreover, whenever possible, the contractor selected the same target locations, on the same side of the vehicle, and with the same horizontal and vertical impact angles as in the compliance tests. Sometimes, however, the design of the pre- and post-standard vehicles were not identical (e.g., due to restyling). If the layout of the pre-standard vehicle made it impossible to test at exactly the same location or angle as in the post-standard vehicle, the contractor chose a nearby location or angle.

---

[15] It is unknown to what extent, if any, the age of the vehicle – specifically, the aging of the foam material at the target locations – might have affected test performance. Ideally, the pre-standard comparison vehicles should have been tested when they were new.

Appendix B documents the parameters for all the (nominally) 24 km/h headform impacts in the pre- and post-standard vehicles, including the actual HIC(d), the target locations, the horizontal and vertical angles and the velocity. When a target location in the pre-standard vehicle was not identical to the one in the compliance test, it shows the closest corresponding location in the compliance test.

Adjusting HIC(d) for impact velocity If the actual impact velocity is slightly lower than 24 km/h, HIC(d) would tend to be lower than at exactly 24 km/h. With the assumptions that the head and what it is striking are both linear systems, HIC would increase by the 2.5 power of the increase in velocity. A 1% increase in velocity would give about a 2.5% increase in HIC.[16] The adjusted HIC(d), the value that would likely have been observed if the test had been run at exactly 24 km/h, is:

$$\text{HIC(d) adjusted} = \text{HIC(d) actual} * (24/\text{SPEED})^{2.5}$$

The adjusted HIC(d) values are also listed in Appendix B and are the ones that will be used in the analyses.

Analysis database Appendix C compares the speed-adjusted HIC(d) in pre- and post-standard vehicles of the same make-model, at identical or closely corresponding target locations. It calculates by how many units HIC(d) was reduced at each location after FMVSS 201. The database contains a total of 154 pairs of HIC(d) readings at matching target locations for the 15 make-models in the study.

Appendix C identifies the actual target locations for the pre- and post-standard impacts and flags the pairs where locations are similar but not exactly the same. Horizontal and vertical impact angles are also compared. Among the 154 pairs, 92 are exceptionally close matches where:

- The pre- and post-standard target locations are identical, and
- The horizontal impact angles in the pre- and post-standard impact differ by no more than 15 degrees, and
- The vertical impact angles also differ by no more than 15 degrees.

Results: overall HIC(d) reduction Compliance testing ultimately measures performance at the vehicle level, on a pass-fail basis only: if HIC(d) exceeds 1000 at a single target, a vehicle fails even if it was far below 1000 at the other targets; if HIC(d) does not exceed 1000 at any target, the vehicle passes, even if all results were barely below 1000. Our main analysis goal, on the other hand, is to <u>quantify</u> the average improvement in HIC(d) after FMVSS 201, target by target (only the last section of the report discusses performance at the vehicle level).

Table 1 shows that HIC(d) averaged 909.9 in 154 head impact tests on pre-standard vehicles, ranging from as low as 426 to as high as 1767. In compliance tests of post-standard vehicles, the 154 impacts to matching locations in the same make-models resulted in a range of HIC(d) from 373 to 986 and an average of 667.5. That is an average improvement of 242.4 units of HIC per test.

---

[16] Eppinger, R. (NHTSA retired) and Willke, D. (NHTSA), e-mails to C.J. Kahane, May 31, 2006.

TABLE 1

AVERAGE HIC(d) BEFORE AND AFTER FMVSS 201
(154 impact locations in 15 make-models)

	Average	Lowest	Highest	Standard Error	t-test
Pre-standard HIC(d)	909.9	426	1767		
Post-standard HIC(d)	667.5	373	986		
HIC(d) improvement	242.4	– 420	1194	24.1	10.06

On the 154 matched pairs of impacts, the improvement in HIC(d) ranged from –420 (i.e., it became worse) to +1194, the average improvement being 242.4. The standard error of the improvement is 24.1. Because t = 242.4/24.1 = 10.06 is much more than the 95th or even the 99th percentile of a t distribution with 153 degrees of freedom, we may reject the null hypothesis that the HIC(d) improvement was zero – i.e., HIC(d) significantly improved, overall, after FMVSS 201.

The preceding error analysis treated the 154 matched pairs as independent observations, a simple random sample. More conservatively, they can be treated as a cluster sample of 15 make-models (primary sampling units), with a second stage of selecting targets within each make-model. The average improvement in HIC(d) is calculated across the various targets in a make-model; then the average improvement in each make-model is averaged across the 15 make-models, weighted by the number of targets in each model. The point estimate of overall average improvement remains 242.4 as in Table 1. In each of the 15 individual make-models, the average of HIC(d) across the various targets for that model improved after FMVSS 201. The average improvement ranged from as low as 90 in one make-model to as high as 534. The standard error for these 15 weighted observations is 30.54.[17] Although t = 242.4/30.54 = 7.94 is somewhat lower than the 10.06 in the preceding analysis, it is still much more than the 99th percentile of a t distribution with 14 degrees of freedom. Here, too, HIC(d) significantly improved after FMVSS 201.

As stated above, the 154 matched pairs included 92 exceptionally close matches. If the other 62 pairs were substantially poorer matches, we would expect the standard deviation of the HIC(d) improvement for these 62 pairs to be substantially higher than for the 92 closest matches (because the pre- and post-standard tests would be less directly comparable). For all 154 pairs, the average HIC improvement is 242.4 and its standard deviation 299. For the 92 closest matches, the average improvement is 260 and its standard deviation 314. For the 62 other pairs, the average improvement is 216 and its standard deviation 277. In other words, the standard deviation for these 62 pairs is quite similar and, in fact, slightly lower than for the 92 exceptionally close matches. The average reduction for the two groups is also reasonably similar (260 vs. 216). The results do not suggest that the 62 pairs are substantially worse matched than the 92. In the remaining analysis, as in Table 1, we may treat all 154 pairs as a single database of well-matched pairs.

---

[17] Calculated by the MEANS procedure of *SAS*®.

Two non-parametric analyses corroborate that HIC(d) scores significantly improved after FMVSS 201. One of them compares the number of HIC(d) scores above 1000 among the pre- and post-standard vehicles. A passing score on the FMVSS 201 compliance test is 1000 or less. Indeed, all 154 head impacts in the compliance tests resulted in HIC(d) less than 1000.[18] Pre-standard vehicles, of course, had no obligation to meet FMVSS 201, and in our tests, 47 of the 154 HIC(d) scores exceeded 1000. The 2x2 table:

	HIC(d) ≤ 1000	HIC(d) > 1000
Pre-FMVSS 201	107	47
Post-FMVSS 201	154	0

has a Chi-square (χ^2) of 55.46. Because χ^2 must exceed 3.84 for statistical significance at the .05 level and 6.64 for significance at the .01 level, the proportion of HIC(d) scores over 1000 is significantly lower after FMVSS 201.[19]

The second analysis examines the 154 matched pairs of impact tests. In 121 of the pairs, HIC(d) was lower on the compliance test than in the corresponding test on the pre-standard vehicle (improvement); in 33 of the pairs it was higher (became worse). That is a statistically significant departure from a 50-50 split:

$$p = 121/154 = .7857; s = \sqrt{(.25/154)} = .0403; z = (p - .5)/s = 7.09$$

HIC(d) reduction by component Table 2 analyzes test results separately for six general regions within vehicles: the A-Pillar, the B-Pillar, other pillars, the upper roof (headliner), roof side rails (including, on minivans, the track above the sliding door, whose location is similar to the roof side rail on other vehicles) and the headers (front and rear). Each general region contains several potential target locations; typically, one or more targets were tested in each of the 15 make-models.

---

[18] Infrequently, NHTSA compliance tests result in failing scores. They trigger a process whereby the manufacturer demonstrates the existing vehicle meets the requirement or modifies the vehicle so it will meet the requirement. However, none of the impacts among our 15 post-standard vehicles resulted in actual HIC(d) over 1000. Additionally, it could happen that actual HIC(d) is 1000 or less, a passing score, but because the impact velocity was slightly less than 24 km/h, our speed adjustment would have pushed HIC(d) over 1000 in our analysis database. This, too, did not occur among our 15 vehicles.

[19] Although one of the actual cell counts in the 2x2 table is 0, a chi-square test is feasible because the expected count in that cell is 23.5. Likewise, Fisher's exact test, the likelihood ratio chi-square, the continuity adjusted chi-square and the Mantel-Haenzel chi-square are all significant at the .01 level.

TABLE 2

AVERAGE HIC(d) BEFORE AND AFTER FMVSS 201, BY COMPONENT

	Average	Lowest	Highest	Standard Error	t-test

A-Pillar (41 impact locations in 15 make-models)

	Average	Lowest	Highest	Standard Error	t-test
Pre-standard HIC(d)	1154.0	455	1767		
Post-standard HIC(d)	678.5	373	986		
HIC(d) improvement	475.5	− 254	1194	52.4	9.07

B-Pillar (31 impact locations in 12 make-models)

Pre-standard HIC(d)	736.7	537	1014		
Post-standard HIC(d)	675.5	489	957		
HIC(d) improvement	61.2	− 420	525	33.9	1.80

Any Other Pillar (25 impact locations in 15 make-models)

Pre-standard HIC(d)	747.1	581	1076		
Post-standard HIC(d)	642.0	464	870		
HIC(d) improvement	105.1	− 179	431	35.4	2.97

Upper Roof (Headliner) (22 impact locations in 15 make-models)

Pre-standard HIC(d)	940.8	543	1363		
Post-standard HIC(d)	662.2	443	872		
HIC(d) improvement	278.7	− 150	739	46.5	5.99

Roof Side Rail (30 impact locations in 15 make-models)

Pre-standard HIC(d)	929.5	617	1481		
Post-standard HIC(d)	684.2	390	898		
HIC(d) improvement	245.3	− 198	806	49.4	4.97

Front or Rear Header (5 impact locations in 4 make-models)

Pre-standard HIC(d)	542.3	426	655		
Post-standard HIC(d)	577.5	414	682		
HIC(d) improvement	− 35.2	− 214	215	72.1	− .49

Performance improvement was largest by far on the A-Pillar. Table 2 shows that HIC(d) averaged 1154.0 in 41 impacts to A-Pillar targets on pre-standard vehicles, ranging as high as 1767. HIC(d) averaged 678.5 in the 41 matching impacts during compliance tests of post-standard vehicles. That is an average improvement of 475.5 units of HIC per test. The standard error of the improvement is 52.4. Because t = 475.5/52.4 = 9.07, HIC(d) significantly improved for the A-pillar after FMVSS 201. The preceding error analysis treated the 41 matched pairs as a simple random sample. Even if they are treated as a cluster sample of 15 make-models (primary sampling units), t = 7.03 is still much more than the 95th percentile of a t distribution with 14 degrees of freedom.

HIC(d) improvement was substantially less at the other pillars. At the B-Pillar, HIC(d) was reduced by an average of 61.2 units in 31 matching pairs of tests on 12 make-models (B-Pillars were not tested on the three pickup trucks in the study). Nevertheless, the value of t is still large enough (in both the simple-random and the cluster-sample approach) that the reduction is statistically significant at the one-sided .05 level. For all other pillars, the average reduction of HIC(d) was 105.1. That, too, is statistically significant.

The interior area of the upper roof (headliner) and the roof side rail were each tested in at least one spot on all 15 make-models. HIC(d) on the upper roof averaged 940.8 before FMVSS 201 and 662.2 afterwards, a statistically significant reduction of 278.7 units. On the roof side rail, HIC(d) improved from 929.5 to 684.2, a statistically significant average reduction of 245.3.

The front or rear headers were tested in only four make-models, at a total of five locations. The front and rear headers had lower HIC(d) scores, on average, before as well as after FMVSS 201, than any other tested component. Average HIC(d) did not improve after FMVSS 201, but these are too few tests for statistically meaningful results.

Table 2 also shows that post-standard performance is quite uniform across the six regions. HIC(d) averages 679 for the A-Pillar, 676 for the B-Pillar, 642 for other pillars, 684 for the roof side rail, 662 for upper roof and 578 for the front and rear header. Moreover, the individual scores are within a tight band and, of course, never exceed 1000. Essentially, FMVSS 201 has set a standard for the entire upper interior of the vehicle. Relative to pre-standard, improvement had to be greatest at the A-pillar because performance was worst there before FMVSS 201 (averaging 1154). By contrast, at the B-Pillar, other pillars and the front/rear header, not so much improvement (if any) was necessary, because performance was usually satisfactory before FMVSS 201, averaging 737, 747 and 542, respectively.

The two non-parametric analyses shed additional light on the comparative effects of FMVSS 201. For each region of the vehicle, a separate 2x2 table compares the number of HIC(d) scores above 1000 among the pre- and post-standard vehicles. The A-Pillar is the only region where the majority of impacts in pre-Standard vehicles (26 of 41) generated HIC(d) over 1000. The 2x2 table for the A-Pillar has a statistically significant χ^2 of 38.07.[20] By contrast, the B-Pillar, other pillars and the front/rear header barely had any scores over 1000 in the pre-standard vehicles. The upper roof and side rail were intermediate, with about one-third of the scores over 1000 before FMVSS 201.

---

[20] Also significant at the .01 level by Fisher's exact test.

		HIC(d) ≤ 1000	HIC(d) > 1000
A-Pillar	Pre-FMVSS 201	15	26
	Post-FMVSS 201	41	0
B-Pillar	Pre-FMVSS 201	30	1
	Post-FMVSS 201	31	0
Other pillars	Pre-FMVSS 201	23	2
	Post-FMVSS 201	25	0
Upper roof	Pre-FMVSS 201	15	7
	Post-FMVSS 201	22	0
Roof side rail	Pre-FMVSS 201	19	11
	Post-FMVSS 201	30	0
Front header	Pre-FMVSS 201	5	0
	Post-FMVSS 201	5	0

The second analysis subdivides the 154 matched pairs of impact tests and identifies how often HIC(d) improved or became worse, post-standard, in each region of the vehicle:

	HIC(d) Improved After FMVSS 201	HIC(d) Became Worse After FMVSS 201
A-Pillar	38	3
B-Pillar	19	12
Any other pillar	18	7
Upper roof (headliner)	20	2
Roof side rail	24	6
Front header	2	3

On the A-Pillar, the upper roof and the roof side rail, the overwhelming majority of targets had better scores in the post-standard vehicle. For each of these three regions, plus "any other pillar," the odds of improvement were significantly better than 50-50. In summary, the nonparametric tests show that scores consistently improved after FMVSS 201 in almost all regions of the car, although on some of those regions they were almost all passing even before FMVSS 201.

It is also interesting to compare the test results with the narrative descriptions, by region of the vehicle, of what parts were added or modified (as shown in the tables earlier in this report). A caveat is that the vehicles in the cost teardown analysis only partially overlapped the test vehicles. Six make-models were cost-analyzed and also tested (Grand Cherokee, Caravan, F-150, Accord, Camry and Sephia); four were cost-analyzed but not tested (Crown Victoria, Taurus, Montana and Jetta); and nine were tested but not cost-analyzed (Neon, Explorer, Windstar, LeSabre, Silverado, Trailblazer, Sentra, Corolla and Tacoma).

The cost analysis showed extensive changes to A-Pillars and the upper roof, consistent with the great improvements in HIC(d). Front and rear headers were little changed, consistent with no improvement in HIC(d). On the other hand, the cost analysis suggested that B-Pillars received many of the same modifications as A-Pillars, except on one vehicle; nevertheless the improvement in HIC(d) was not nearly as great for the B-Pillar as the A-Pillar. Conversely, the cost analysis said other pillars and the roof side rail were unchanged on most vehicles, but the tests showed significant improvements in HIC(d). These inconsistencies are not easily explained. A closer look at the cost analysis perhaps suggests that the B-Pillar modifications, although qualitatively similar to those on the A-Pillar, are often less extensive. Perhaps the test improvements on the roof side rail are attributable to the modifications to the upper roof and the pillars in the vicinity of the side rail.

HIC(d) reduction by vehicle type Table 3 shows that each of the four types of vehicles in our study sample – passenger cars, pickup trucks, SUVs and minivans – experienced a statistically significant reduction of average HIC(d) after FMVSS 201. Before the standard, average HIC(d) was slightly worse in LTVs (955.6) than in passenger cars (855.1); however, the improvement was slightly greater in the LTVs. After FMVSS 201, all four vehicle types have HIC(d) scores on compliance tests averaging in the mid-600s.

TABLE 3

AVERAGE HIC(d) BEFORE AND AFTER FMVSS 201, BY VEHICLE TYPE

	Average HIC(d)			t-test for $\Delta = 0$
	Pre-Standard	Post-Standard	Δ	
Passenger cars	855.1	682.8	172.3	6.00
All LTVs	955.6	654.7	300.9	8.34
Pickup trucks	912.8	627.6	285.2	4.83
SUVs	965.4	669.9	295.5	6.25
Minivans	988.3	664.6	323.7	3.88

Performance at the vehicle level All 17 of the post-standard vehicles in our study met the compliance test for FMVSS 201 because they had actual HIC(d) scores of 1000 or less at every target tested by NHTSA.[21] By contrast, the results in Appendix B show that only 2 of the 15 pre-standard vehicles had actual HIC(d) ≤ 1000 on each target – and on one of these vehicles, HIC(d) might have exceeded 1000 at one target if the test had been conducted at 24.0 km/h rather than the actual 23.7 km/h. In other words, after FMVSS 201, performance greatly improved at the overall vehicle level as well as on a target-by-target basis.

Conclusions The 15 make-models tested in this analysis were purposively selected from the 68 vehicles compliance-tested in FY 1999-2003. They are not a probability sample of all the make-models of passenger vehicles. Thus, the analysis findings pertain to these 15 models that comprise nearly a quarter of all new-vehicle sales and include a variety of manufacturers and vehicle types (passenger cars, pickup trucks, SUVs and minivans) – but they do not necessarily generalize to all make-models subject to the FMVSS 201 upgrade.

The preceding analyses demonstrate a substantial reduction of HIC(d) after the FMVSS 201 upgrade for those 15 high-sales make-models. HIC has been validated as a criterion for predicting head injury risk. However, this study does not analyze crash data to quantify the actual fatality- or injury-reducing effectiveness of the equipment introduced in 1999-2003 to upgrade head impact protection. That evaluation will be a high priority for NHTSA during the next four years.[22]

---

[21] The study comprised 15 make-models, but compliance tests have been performed on two post-standard Honda Accords and two Toyota Camrys – see Appendix B.

[22] *National Highway Traffic Safety Administration Evaluation Program Plan, Calendar Years 2004-2007*, p. 8.

REFERENCES

Campbell, B.J. *A Study of Injuries Related to Padding on Instrument Panels*. HSL Publication No. 00427812, Report No. VJ-1823-R2, Cornell Aeronautical Laboratory, Buffalo, 1963.

Code of Federal Regulations, Title 49. Government Printing Office, Washington, 2006.

Federal Register Notices:

> 31 (December 3, 1966): 15212, Notice of Proposed Rulemaking for the initial Federal Motor Vehicle Safety Standards, including FMVSS 201.

> 60 (August 18, 1995): 43031, Final Rule upgrading FMVSS 201.

> 63 (August 4, 1998): 41451, Final Rule reducing the FMVSS 201 test speed from 15 mph to 12 mph on target areas where a head-protection air bag is stored.

Final Economic Assessment, FMVSS No. 201, Upper Interior Head Protection. NHTSA Plans and Policy, Washington, 1995.

Kahane, C.J. *An Evaluation of Occupant Protection in Frontal Interior Impact for Unrestrained Front Seat Occupants of Cars and Light Trucks*. NHTSA Technical Report No. DOT HS 807 203, Washington, 1988.

_____. *An Evaluation of Side Impact Protection – FMVSS 214 TTI(d) Improvements and Side Air Bags*. NHTSA Technical Report, to appear in 2007.

Laboratory Test Procedure for FMVSS 201U – Occupant Protection in Interior Impact – Upper Interior Head Impact Protection. Procedure No. TP201U-01, NHTSA Office of Vehicle Safety Compliance, Washington, April 3,1998, accessible from http://nhtsa.gov/portal/site/nhtsa/menuitem.b166d5602714f9a73baf3210dba046a0/.

Ludtke, N.F., Osen, W., Gladstone, R., and Lieberman, W. *Perform Cost and Weight Analysis, Non Air Bag Head Protection Systems, FMVSS 201*. NHTSA Technical Report No. DOT HS 809 810, Washington, 2003.

National Highway Traffic Safety Administration Evaluation Program Plan, Calendar Years 2004-2007. NHTSA Report No. DOT HS 809 699, Washington, 2004.

SAE Handbook. Society of Automotive Engineers, Warrendale, PA, annual publication.

APPENDIX A

NHTSA's SPECIFICATIONS FOR USED VEHICLES
FOR HEAD-IMPACT TESTING

The vehicle must be totally complete. Mileage and cosmetics are not important. These cars [sic] will be used to test head impact protection in pre-standard vehicles. Head impacts with the upper interior components including pillars, side rails, headers, and the roof will be conducted. Structural soundness of the body is most important. The vehicle must be able to run and drive. Specifically:

- **Body**. All areas of the vehicle that are potential targets in the FMVSS 201 test, and any surrounding areas, must be intact on the interior and exterior of the vehicle. The upper interior components must be completely undamaged and composed of the original trim. The vehicle shall not have been in a crash except for minor front-end or rear-end damage (broken grille, headlamps, turn signal lenses, etc.). Damaged bumpers are acceptable. The condition of the paint is unimportant. The doors must open and close properly and be correctly aligned so that there are no gaps when closed.

- **Drive train**. The engine, transmission, and drive axles must be present and in their proper place. The engine, transmission, and parking brake must operate properly. All under-the-hood components must be in place (battery, alternator, radiator, etc.). The battery must hold a charge so that electrical accessories in the vehicle may be operated without starting the engine. The radiator and its supporting cross members must be undamaged. The cooling system must be able to retain its coolant (small leaks are acceptable), and the proper fluid levels must be maintained. The fuel tank must not leak and must be completely undamaged with no dents that would reduce its internal capacity.

- **Interior**. The interior of the vehicle must be complete with properly functioning seats and seatbelts. The interior trim pieces must be in their proper place. The front seat must slide freely forward and backward to allow adjustment. The seat backs must be able to be adjusted freely. Head restraints must be in place and must adjust freely (if adjustable, some are not). All the windows in the vehicle must operate properly. The door latches and locks (power or manual) must operate properly.

- **Suspension**. The vehicle must not sag or have a raised/lowered suspension. Original condition is very important. Some slight sagging due to age is expected, but no one side or corner of the vehicle should be sagging. None of the control arms or other suspension components shall be damaged or distorted. Tires must have some usable tread remaining and hold pressure.

APPENDIX B

RESULTS OF COMPLIANCE TESTS FOR 15 MAKE-MODELS AND
HEAD IMPACT TESTS ON CORRESPONDING PRE-STANDARD VEHICLES

Column 1   Name of the make-model preceded by the 5-digit make-model code used in NHTSA evaluations

Column 2   Model year

Column 3   Test vehicle number (6-character code for the specimen test vehicle, may be used to find a detailed report in the NHTSA test database.

Columns 4-6   Identifies the target locations tested. Column 4 is the actual target location. Column 5 is the target location that will be treated in the analysis as equivalent or corresponding to the actual location. Column 6 flags with a * when the actual and equivalent locations are not the same. (AP = A-Pillar, BP = B-Pillar, OP = other pillar, RP = rear pillar, UR = upper roof, SR = roof side rail, SD = upper sliding door track on minivan, FH = front header, RH = rear header)

Columns 7-8   Horizontal and vertical impact angles. Horizontal impact angle measures degrees away from straight-ahead. Vertical angle measures degrees away from level (positive = upwards, negative = downwards).

Column 9   Test speed in km/h. The standard specifies 15 mph tests; that is approximately 24 km/h. When a head-protection air bag is situated beneath the target location, the standard specifies 12 mph tests; these are excluded from the analysis and not shown in Appendices B and C.

Columns 10-11   HIC recorded on the dummy. Column 10 is actual HIC. Column 11 is HIC adjusted up or down to a 24 km/h impact speed = (24/SPD) x actual HIC.

MAKE-MODEL & CODE	MODEL YEAR	TEST VEH NO.	TARGET ACTUAL	TARGET EQUIV TO	DIFFERENT?	IMPACT ANGLE H	IMPACT ANGLE V	TEST SPD km/h	HIC(d) ACTUAL	HIC(d) ADJ
COMPLIANCE TEST:										
2312 JEEP GRAND CHEROKEE	1999	CX0307	AP1	AP1		110	29	23.8	733	748
			AP2	AP2		202	46	23.8	867	885
			AP3	AP3		160	40	23.5	687	724
			BP3	BP3		80	-8	23.7	579	597
			BP4	BP4		270	-9	24.0	679	679
			OP1	OP1		270	8	24.1	514	509
			OP2	OP2		90	-6	24.0	505	505
			SR1	SR1		270	32	24.0	674	674
			SR2A	SR2A		270	33	24.0	804	804
			UR4	UR4		90	41	24.0	872	872
EQUIVALENT PRE-STANDARD VEHICLE:										
2312 JEEP GRAND CHEROKEE	1996	PT0400	AP1	AP1		105	31	23.5	1319	1390
			AP2	AP2		201	41	23.6	1309	1365
			AP3	AP3		160	39	23.3	1348	1452
			BP3	BP3		90	-3	23.5	648	683
			BP4	BP4		215	-6	23.5	734	774
			OP1	OP1		270	6	24.0	798	798
			OP2	OP2		90	1	23.7	816	842
			SR1	SR1		270	35	23.8	1412	1442
			SR2A	SR2A		270	33	23.7	1061	1095
			UR4	UR4		90	50	23.8	1131	1155

NOTE: THE POST-STANDARD AND PRE-STANDARD VEHICLE WERE TESTED AT THE SAME TARGET LOCATIONS.

| | MODEL YEAR | TEST VEH NO. | TARGET ACTUAL | TARGET EQUIV TO | TARGET DIFFERENT? | IMPACT ANGLE H | IMPACT ANGLE V | TEST SPD km/h | HIC(d) ACTUAL | HIC(d) ADJ |
MAKE-MODEL & CODE										
COMPLIANCE TEST:										
7020 DODGE NEON	2000	CY0305	AP1	AP1		144	41	23.8	593	606
			AP2	AP2		199	50	23.4	668	712
			AP3	AP3		161	48	23.7	874	902
			BP1	BP1		90	22	23.9	651	658
			BP3	BP3		286	-2	24.2	590	578
			BP4	BP4		150	-6	23.2	879	957
			RP2	RP2		70	20	23.8	605	618
			SR1	SR1		270	43	24.3	725	703
			SR2A	SR2A		90	43	23.9	801	809
			UR2	UR2		270	28	23.7	698	720
EQUIVALENT PRE-STANDARD VEHICLE:										
7020 DODGE NEON	1997	PV0300	AP1	AP1		130	42	23.6	793	827
			AP2	AP2		205	44	23.4	926	987
			AP3	AP3		154	41	23.9	1136	1148
			BP1	BP1		90	11	23.9	885	894
			BP3	BP3		280	2	23.6	591	616
			BP4	BP4		157	-6	23.6	515	537
			RP2	RP2		40	21	23.6	853	890
			SR1	SR1		270	33	23.5	1085	1144
			SR2A	SR2A		90	27	23.1	897	987
			UR2	UR2		270	25	23.3	980	1055

NOTE: THE POST-STANDARD AND PRE-STANDARD VEHICLE WERE TESTED AT THE SAME TARGET LOCATIONS.

MAKE-MODEL & CODE	MODEL YEAR	TEST VEH NO.	TARGET ACTUAL	TARGET EQUIV TO	TARGET DIFFERENT?	IMPACT ANGLE H	IMPACT ANGLE V	TEST SPD km/h	HIC(d) ACTUAL	HIC(d) ADJ
COMPLIANCE TEST:										
7406 DODGE GRAND CARAVAN	2002	C20309	AP1	AP1		140	28	23.9	567	573
			AP2	AP2		201	47	23.9	707	714
			AP3	AP3		160	44	23.9	510	515
			BP2	BP2		90	6	23.5	595	627
			BP3	BP3		270	0	24.0	679	679
			OP2	OP2		90	-4	23.5	658	694
			SD	SD		90	47	23.7	580	599
			SR1	SR1		270	32	23.7	654	675
			UR2	UR2		270	45	23.5	446	470
			UR4	UR4		90	32	23.8	611	624
EQUIVALENT PRE-STANDARD VEHICLE:										
7406 DODGE GRAND CARAVAN	1996	PTO301	AP1	AP1		115	50	23.9	1749	1767
			AP2	AP2		205	46	23.9	1417	1432
			AP3	AP3		157	42	23.9	1597	1614
			BP2	BP2		90	13	23.7	811	837
			BP3	BP3		270	5	23.5	707	745
			OP2	OP2		270	3	24.0	716	716
			SD	SD		90	50	23.0	694	772
			SR1	SR1		270	28	23.9	1466	1481
			UR2	UR2		270	50	24.1	791	783
			UR4	UR4		90	49	23.8	1335	1363

NOTE: THE POST-STANDARD AND PRE-STANDARD VEHICLE WERE TESTED AT THE SAME TARGET LOCATIONS.

| MAKE-MODEL & CODE | MODEL YEAR | TEST VEH NO. | T A R G E T | | | IMPACT ANGLE | | TEST SPD | HIC(d) | |
			ACTUAL	EQUIV TO	DIFFERENT?	H	V	km/h	ACTUAL	ADJ
COMPLIANCE TEST:										
12212 FORD F150 SUPERCAB PK	2002	C20210	AP1	AP1		141	8	23.9	830	839
			AP2	AP2		202	30	23.8	577	589
			AP3	AP3		150	41	23.9	701	708
			RH	RH		0	50	23.5	623	657
			RP2	RP2		85	8	23.9	611	617
			SR2A	SR2A		270	32	23.9	507	512
			SR3	SR3		90	35	23.5	839	884
			UR1	UR1		270	50	23.8	719	734
			UR2	UR2		270	38	23.9	636	643
			UR6	UR6		45	50	23.8	640	654
EQUIVALENT PRE-STANDARD VEHICLE:										
12212 FORD F150 SUPERCAB PK	1998	PW0201	AP1	AP1		110	31	23.7	1182	1220
			AP2	AP2		201	38	23.5	1331	1403
			AP3	AP3		158	38	23.6	436	455
			RH	RH		0	40	23.5	420	443
			RP2	RP2		270	0	23.5	670	706
			SR2A	SR2A		270	35	23.5	1039	1095
			SR31	SR3	*	270	24	23.9	679	686
			UR1	UR1		270	50	23.6	1211	1263
			UR2	UR2		270	50	23.4	686	731
			UR3	UR6	*	270	50	23.7	962	993

NOTE: IN THE ANALYSIS, RESULTS FOR LOCATIONS SR31 AND UR3 ON THE PRE-STANDARD VEHICLE WILL BE COMPARED TO SR3 AND UR6, RESPECTIVELY, ON THE COMPLIANCE TEST.

31

MAKE-MODEL & CODE	MODEL YEAR	TEST VEH NO.	TARGET			IMPACT ANGLE		TEST SPD km/h	HIC(d)	
			ACTUAL	EQUIV TO	DIFFER ENT?	H	V		ACTUAL	ADJ

COMPLIANCE TEST:

MAKE-MODEL & CODE	MODEL YEAR	TEST VEH NO.	ACTUAL	EQUIV TO	DIFFERENT?	H	V	TEST SPD km/h	ACTUAL	ADJ
12302 FORD EXPLORER 4DR	2002	C20200	AP1	AP1		130	38	24.3	870	843
			AP2	AP2		219	40	23.5	764	805
			AP3	AP3		134	24	23.8	380	388
			BP1	BP1		90	6	23.7	715	738
			BP2	BP2		270	9	23.8	849	867
			BP4	BP4		143	-10	23.8	534	545
			FH2	FH2		90	50	23.9	477	482
			OP1	OP1		90	5	23.4	462	492
			SR1	SR1		270	39	23.9	805	813
			UR1	UR1		270	50	23.9	438	443
			UR2	UR2		360	50	23.4	447	476

EQUIVALENT PRE-STANDARD VEHICLE:

MAKE-MODEL & CODE	MODEL YEAR	TEST VEH NO.	ACTUAL	EQUIV TO	DIFFERENT?	H	V	TEST SPD km/h	ACTUAL	ADJ
12302 FORD EXPLORER 4DR	1998	PW0202	AP1	AP1		110	33	23.8	884	903
			AP2	AP2		205	40	23.7	865	893
			AP3	AP3		155	35	23.8	895	914
			BP1	BP1		90	11	23.7	966	997
			BP2	BP2		270	4	23.8	655	669
			BP4	BP4		140	-5	23.7	520	537
			FH2	FH2		180	50	24.0	426	426
			OP1	OP1		90	1	23.5	553	583
			SR1	SR1		270	31	23.6	1174	1224
			UR1	UR1		270	50	23.5	1083	1142
			UR2	UR2		270	50	23.7	928	958

NOTE: THE POST-STANDARD AND PRE-STANDARD VEHICLE WERE TESTED AT THE SAME TARGET LOCATIONS.

32

COMPLIANCE TEST:

MAKE-MODEL & CODE	MODEL YEAR	TEST VEH NO.	TARGET			IMPACT ANGLE		TEST SPD	HIC(d)	
			ACTUAL	EQUIV TO	DIFFERENT?	H	V	km/h	ACTUAL	ADJ
12402 FORD WINDSTAR WAGON	1999	CX0208	AP1	AP1		147	25	23.6	682	711
			AP2	AP2		213	42	23.4	457	487
			AP3	AP3		144	35	23.9	645	652
			BP1	BP1		270	22	23.6	469	489
			BP2	BP2		90	5	23.7	742	766
			BP4	BP4		270	70	23.6	684	713
			FH1	FH1		180	50	24.7	701	652
			OP1	OP1		90	-3	23.3	592	637
			OP2	OP2		270	5	23.3	635	684
			RP1	RP1		317	50	23.6	686	715
			RP2	RP2		90	45	23.7	843	870
			SD	SD		270	30	23.4	655	698
			SR1	SR1		270	40	23.5	852	898
			SR2A	SR2A		90	45	23.7	843	870
			SR3	SR3		90	50	23.9	756	764
			UR	UR		90	38	23.9	498	503

EQUIVALENT PRE-STANDARD VEHICLE:

MAKE-MODEL & CODE	MODEL YEAR	TEST VEH NO.	TARGET			IMPACT ANGLE		TEST SPD	HIC(d)	
			ACTUAL	EQUIV TO	DIFFERENT?	H	V	km/h	ACTUAL	ADJ
12402 FORD WINDSTAR WAGON	1998	PWO200	AP1	AP1		155	43	23.8	994	1015
			AP2	AP2		205	46	23.8	1420	1450
			AP3	AP3		155	40	24.0	1702	1702
			BP1	BP1		90	21	24.0	1014	1014
			BP2	BP2		90	0	23.9	738	746
			BP4	BP4		207	-8	23.9	594	600
			FH1	FH1		180	50	24.0	655	655
			OP1	OP1		270	0	23.9	690	697
			OP2	OP2		90	-1	23.6	578	603
			RP1	RP1		40	44	23.9	772	780
			RP2	RP2		345	13	24.5	727	690
			SD	SD		90	28	24.0	825	825
			SR1	SR1		270	38	23.8	866	884
			SR2A	SR2A		270	35	23.7	767	792
			SR31	SR3	*	270	39	23.5	696	734
			UR2	UR	*	270	41	23.8	978	999

NOTE: IN THE ANALYSIS, RESULTS FOR LOCATIONS SR31 AND UR2 ON THE PRE-STANDARD VEHICLE WILL BE COMPARED TO SR3 AND UR, RESPECTIVELY, ON THE COMPLIANCE TEST.

MAKE-MODEL & CODE	MODEL YEAR	TEST VEH NO.	TARGET ACTUAL	TARGET EQUIV TO	TARGET DIFFERENT?	IMPACT ANGLE H	IMPACT ANGLE V	TEST SPD km/h	HIC(d) ACTUAL	HIC(d) ADJ
COMPLIANCE TEST:										
18002 BUICK LeSABRE	2001	C10110	AP1	AP1		250	30	23.9	667	674
			AP3	AP3		139	39	23.9	703	710
			BP1	BP1		90	20	23.8	679	693
			BP3	BP3		281	-6	23.9	811	820
			BP4	BP4		125	-5	23.6	627	654
			RP1	RP1		285	29	23.7	823	849
			RP2	RP2		46	26	23.5	672	708
			SR2B	SR2B		90	50	24.2	785	769
			SR3	SR3		270	50	23.5	695	733
			UR	UR		270	50	23.9	656	663
EQUIVALENT PRE-STANDARD VEHICLE:										
18002 BUICK LeSABRE	1998	PW0100	AP1	AP1		250	38	23.9	1064	1075
			AP3	AP3		149	43	23.6	1296	1352
			BP1	BP1		90	5	23.9	598	604
			BP3	BP3		275	3	24.0	575	575
			BP4	BP4		161	-4	24.0	547	547
			RP1	RP1		290	30	23.6	692	722
			RP2	RP2		67	23	23.9	575	581
			SR2B	SR2B		90	45	23.8	1051	1073
			SR31	SR3	*	270	31	23.8	734	750
			UR2	UR	*	270	37	23.7	869	897

NOTE: IN THE ANALYSIS, RESULTS FOR LOCATIONS SR31 AND UR2 ON THE PRE-STANDARD VEHICLE WILL BE COMPARED TO SR3 AND UR, RESPECTIVELY, ON THE COMPLIANCE TEST.

MAKE-MODEL & CODE	MODEL YEAR	TEST VEH NO.	TARGET			IMPACT ANGLE		TEST SPD km/h	HIC(d)	
			ACTUAL	EQUIV TO	DIFFERENT?	H	V		ACTUAL	ADJ

COMPLIANCE TEST:

MAKE-MODEL & CODE	MODEL YEAR	TEST VEH NO.	ACTUAL	EQUIV TO	DIFFERENT?	H	V	TEST SPD km/h	ACTUAL	ADJ
20212 CHEVY SILVERADO X-CAB PK	1999	CX0115	AP1	AP1		249	25	23.6	639	666
			AP2	AP2		119	35	23.8	452	462
			AP3	AP3		216	32	23.8	540	551
			RP1	RP1		274	22	23.8	727	742
			RP2	RP2		90	10	23.3	621	669
			SR2A	SR2A		270	48	23.3	649	699
			UR	UR		270	50	23.5	658	694

EQUIVALENT PRE-STANDARD VEHICLE:

MAKE-MODEL & CODE	MODEL YEAR	TEST VEH NO.	ACTUAL	EQUIV TO	DIFFERENT?	H	V	TEST SPD km/h	ACTUAL	ADJ
20212 CHEVY C-1500 X-CAB PK	1998	PW0102	AP1	AP1		248	35	24.5	1377	1308
			AP2	AP2		161	29	24.2	1241	1216
			AP3	AP3		201	32	23.8	1035	1057
			RP1	RP1		270	7	23.8	865	883
			RP2	RP2		90	5	23.2	534	581
			SR2A	SR2A		270	23	23.8	1039	1061
			UR2	UR	*	270	50	23.8	532	543

NOTE: IN THE ANALYSIS, THE RESULT FOR LOCATION UR2 ON THE PRE-STANDARD VEHICLE WILL BE COMPARED TO UR ON THE COMPLIANCE TEST.

MAKE-MODEL & CODE	MODEL YEAR	TEST VEH NO.	T A R G E T			IMPACT A N G L E		TEST SPD	HIC(d)	
			ACTUAL	EQUIV TO	DIFFER ENT?	H	V	km/h	ACTUAL	ADJ

COMPLIANCE TEST:

MAKE-MODEL & CODE	MODEL YEAR	TEST VEH NO.	ACTUAL	EQUIV TO	DIFFER ENT?	H	V	km/h	ACTUAL	ADJ
20302 CHEVY TRAILBLAZER 4DR	2002	C20106	AP1	AP1		246	38	23.5	571	602
			AP2	AP2		143	40	23.3	729	785
			AP3	AP3		214	36	23.7	955	986
			BP1	BP1		90	21	23.3	676	728
			BP3	BP3		281	-6	23.6	850	886
			BP4	BP4		122	-10	23.3	501	539
			OP1	OP1		90	-10	23.9	493	498
			SR2A	SR2A		270	50	23.6	684	713
			SR3	SR3		90	50	23.3	506	545
			UR	UR		270	50	23.6	569	593

EQUIVALENT PRE-STANDARD VEHICLE:

MAKE-MODEL & CODE	MODEL YEAR	TEST VEH NO.	ACTUAL	EQUIV TO	DIFFER ENT?	H	V	km/h	ACTUAL	ADJ
20302 CHEVY BLAZER 4DR	1996	PT0101	AP1	AP1		113	25	23.7	1169	1206
			AP2	AP2		206	38	23.9	1293	1307
			AP3	AP3		155	36	24.2	1497	1466
			BP1	BP1		90	27	23.8	669	683
			BP3	BP3		270	-2	23.5	753	794
			BP4	BP4		110	-10	23.4	785	836
			OP1	OP1		90	5	23.9	745	753
			SR2A	SR2A		270	43	23.6	1199	1250
			SR33	SR3	*	90	30	24.0	665	665
			UR2	UR	*	270	50	23.6	697	727

NOTE: IN THE ANALYSIS, RESULTS FOR LOCATIONS SR33 AND UR2 ON THE PRE-STANDARD VEHICLE WILL BE COMPARED TO SR3 AND UR, RESPECTIVELY, ON THE COMPLIANCE TEST.

36

MAKE-MODEL & CODE	MODEL YEAR	TEST VEH NO.	TARGET ACTUAL	TARGET EQUIV TO	TARGET DIFFERENT?	IMPACT ANGLE H	IMPACT ANGLE V	TEST SPD km/h	HIC(d) ACTUAL	HIC(d) ADJ
COMPLIANCE TEST:										
35043 NISSAN SENTRA	2001	C15202	AP1	AP1		125	32	23.5	829	874
			AP2	AP2		201	48	23.7	596	615
			AP3	AP3		159	44	23.1	507	558
			BP1	BP1		90	22	23.8	747	763
			BP2	BP2		90	11	23.9	678	685
			BP3	BP3		279	-3	23.8	579	591
			RP1	RP1		90	15	23.6	552	576
			SR1	SR1		270	46	23.4	711	757
			SR2A	SR2A		270	37	23.9	821	830
			UR	UR		270	36	24.0	832	832
EQUIVALENT PRE-STANDARD VEHICLE:										
35043 NISSAN SENTRA	1998	PW5200	AP1	AP1		113	21	23.8	907	926
			AP2	AP2		203	47	23.8	1284	1311
			AP3	AP3		157	44	24.0	1584	1584
			BP1	BP1		90	12	23.8	742	758
			BP2	BP2		90	12	24.4	670	643
			BP3	BP3		284	-4	23.9	770	778
			RP1	RP1		90	14	23.6	632	659
			SR1	SR1		270	20	23.9	875	884
			SR2A	SR2A		270	18	24.0	689	689
			UR2	UR	*	270	38	24.0	955	955

NOTE: IN THE ANALYSIS, THE RESULT FOR LOCATION UR2 ON THE PRE-STANDARD VEHICLE WILL BE COMPARED TO UR ON THE COMPLIANCE TEST.

37

| | | | T A R G E T | | | IMPACT A N G L E | | TEST SPD | HIC(d) | |
MAKE-MODEL & CODE	MODEL YEAR	TEST VEH NO.	ACTUAL	EQUIV TO	DIFFER ENT?	H	V	km/h	ACTUAL	ADJ
COMPLIANCE TESTS:										
37032 HONDA ACCORD	1999	CX5303	AP2	AP2		246	40	23.2	533	580
			BP2	BP2		90	10	23.6	546	569
			RP2	RP2		70	20	24.1	704	697
			SR2	SR2		90	50	24.1	538	532
37032 HONDA ACCORD	2003	C35301	BP2	BP2		90	22	23.8	719	734
			RP2	RP2		70	23	23.9	620	627
			UR2	UR2		90	36	23.4	718	765
			UR3	UR3		90	50	23.8	495	505
			UR4	UR4		90	43	23.5	590	622
			UR5	UR5		180	48	24.0	630	630
EQUIVALENT PRE-STANDARD VEHICLE:										
37032 HONDA ACCORD	1996	PT5300	AP2	AP2		206	50	23.9	1036	1047
			BP2	BP2		90	20	23.7	804	830
			RP2	RP2		80	8	24.0	600	600
			SR2A	SR2	*	270	29	24.0	1248	1248
			UR2	UR2		270	38	23.8	741	757
			UR3	UR3		270	50	23.8	886	905
			UR4	UR4		90	46	23.9	832	841
			UR5	UR5		90	39	23.9	816	825

NOTE: IN THE ANALYSIS, THE RESULTS FOR LOCATION BP2 ON THE 1999 AND 2003 COMPLIANCE TESTS WILL BE AVERAGED; SIMILARLY, FOR RP2. THE RESULT FOR LOCATION SR2A ON THE PRE-STANDARD VEHICLE WILL BE COMPARED TO SR2 ON THE 1999 COMPLIANCE TEST.

			T A R G E T			IMPACT ANGLE		TEST SPD	HIC(d)	
MAKE-MODEL & CODE	MODEL YEAR	TEST VEH NO.	ACTUAL	EQUIV TO	DIFFERENT?	H	V	km/h	ACTUAL	ADJ
COMPLIANCE TEST:										
49032 TOYOTA COROLLA	2003	C35101	AP1	AP1		124	17	23.6	674	703
			AP3	AP3		155	50	23.6	554	578
			BP1	BP1		90	12	23.6	611	637
			BP2	BP2		270	4	23.5	555	585
			BP3	BP3		88	-8	23.7	496	512
			RP1	RP1		60	16	23.5	525	553
			RP2	RP2		296	17	23.5	548	578
			SR1	SR1		270	22	23.5	552	582
			SR2A	SR2A		90	10	23.6	374	390
			UR	UR		270	46	23.7	736	760
EQUIVALENT PRE-STANDARD VEHICLE:										
49032 TOYOTA COROLLA	1998	PW5100	AP1	AP1		109	21	23.7	604	623
			AP3	AP3		157	42	23.7	627	647
			BP1	BP1		90	10	24.1	702	695
			BP2	BP2		270	8	23.4	553	589
			BP3	BP3		83	-1	23.7	591	610
			RP1	RP1		82	18	23.9	656	663
			RP2	RP2		280	13	23.9	764	772
			SR1	SR1		270	27	24.5	880	836
			SR2A	SR2A		90	17	23.7	612	632
			UR2	UR	*	270	37	23.6	782	816

NOTE: IN THE ANALYSIS, THE RESULT FOR LOCATION UR2 ON THE PRE-STANDARD VEHICLE WILL BE COMPARED TO UR ON THE COMPLIANCE TEST.

COMPLIANCE TESTS:

MAKE-MODEL & CODE	MODEL YEAR	TEST VEH NO.	TARGET ACTUAL	TARGET EQUIV TO	TARGET DIFFERENT?	IMPACT ANGLE H	IMPACT ANGLE V	TEST SPD km/h	HIC(d) ACTUAL	HIC(d) ADJ
49040 TOYOTA CAMRY	1999	CX5107	AP1	AP1		115	30	23.9	930	940
			AP2	AP2		221	45	23.5	733	773
			AP3	AP3		121	40	23.5	752	793
			BP1	BP1		270	20	23.9	720	728
			BP4	BP4		118	0	23.3	691	744
			RP1	RP1		286	35	24.0	641	641
			RP2	RP2		71	37	23.7	941	971
			SR2A	SR2A		270	30	23.8	863	881
			SR3	SR32	*	90	32	24.6	682	641
49040 TOYOTA CAMRY	2002	C25101	AP1	AP1		111	20	23.5	540	569
			AP2	AP2		221	48	23.9	634	641
			AP3	AP3		145	46	23.8	491	501
			BP1	BP1		270	8	23.9	533	539
			BP2	BP2		270	11	23.5	704	742
			BP4	BP4		110	-8	23.5	652	687
			RP1	RP1		295	11	23.3	541	583
			RP2	RP2		74	27	23.4	666	710
			SR3	SR31	*	270	35	23.5	483	509
			UR	UR		90	45	23.5	819	863

NOTE 1: IN THE ANALYSIS, THE RESULTS FOR LOCATION AP1 ON THE 1999 AND 2002 COMPLIANCE TESTS WILL BE AVERAGED; SIMILARLY FOR AP2, AP3, BP1, BP4 AND RP2.

EQUIVALENT PRE-STANDARD VEHICLE:

MAKE-MODEL & CODE	MODEL YEAR	TEST VEH NO.	TARGET ACTUAL	TARGET EQUIV TO	TARGET DIFFERENT?	IMPACT ANGLE H	IMPACT ANGLE V	TEST SPD km/h	HIC(d) ACTUAL	HIC(d) ADJ
49040 TOYOTA CAMRY	1998	PW5101	AP1	AP1		108	32	23.5	714	753
			AP2	AP2		204	47	23.7	907	936
			AP3	AP3		157	43	23.7	959	990
			BP1	BP1		270	24	23.8	931	951
			BP2	BP2		270	11	24.0	922	922
			BP4	BP4		160	-8	23.8	713	728
			RP1	RP1		270	22	23.7	620	640
			RP2	RP2		70	21	24.0	730	730
			SR2A	SR2A		270	28	23.9	871	880
			SR31	SR31		270	34	23.9	619	625
			SR32	SR32		90	38	23.8	650	664
			UR5	UR	*	90	39	23.7	981	1012

NOTE2: THE RESULT FOR UR6 ON THE PRE-STANDARD VEHICLE WILL BE COMPARED TO UR ON THE 2002 COMPLIANCE TEST; LIKEWISE SR31 (270 HORIZONTAL) TO SR3 (270 HORIZONTAL) IN 2002; SR32 (90 HORIZONTAL) TO SR3 (90 HORIZONTAL) IN 1999.

COMPLIANCE TEST:

MAKE-MODEL & CODE	MODEL YEAR	TEST VEH NO.	TARGET			IMPACT ANGLE		TEST SPD	HIC(d)	
			ACTUAL	EQUIV TO	DIFFERENT?	H	V	km/h	ACTUAL	ADJ
49202 TOYOTA TACOMA XTRACAB	2003	C35102	AP1	AP1		150	20	23.7	577	595
			AP2	AP2		230	30	23.6	523	545
			AP3	AP3		130	30	23.5	354	373
			FH1	FH1		180	50	23.6	654	682
			RH	RH		0	34	23.4	389	414
			RP1	RP1		52	17	23.4	589	627
			RP2	RP2		78	1	23.7	693	715
			SR1	SR1		270	25	23.4	401	427
			SR2	SR2		90	31	23.4	481	512
			UR	UR		276	44	23.8	716	731

EQUIVALENT PRE-STANDARD VEHICLE:

MAKE-MODEL & CODE	MODEL YEAR	TEST VEH NO.	TARGET			IMPACT ANGLE		TEST SPD	HIC(d)	
			ACTUAL	EQUIV TO	DIFFERENT?	H	V	km/h	ACTUAL	ADJ
49202 TOYOTA TACOMA XTRACAB	1997	PV5102	AP1	AP1		137	35	23.7	1086	1121
			AP2	AP2		217	30	23.8	912	931
			AP3	AP3		144	30	23.7	685	707
			FH1	FH1		180	50	23.6	536	559
			RH	RH		0	31	23.8	616	629
			RP1	RP1		45	29	23.8	1037	1059
			RP2	RP2		270	5	23.3	999	1076
			SR1	SR1		270	33	24.0	1086	1086
			SR2A	SR2	*	270	31	23.9	797	805
			UR5	UR	*	90	50	23.5	976	1029

NOTE: IN THE ANALYSIS, RESULTS FOR LOCATIONS SR2A AND UR5 ON THE PRE-STANDARD VEHICLE WILL BE COMPARED TO SR2 AND UR, RESPECTIVELY, ON THE COMPLIANCE TEST.

41

	MAKE-MODEL & CODE	MODEL YEAR	TEST VEH NO.	ACTUAL	EQUIV TO	DIFFERENT?	ANGLE H	ANGLE V	TEST SPD km/h	ACTUAL	ADJ

COMPLIANCE TEST:

MAKE-MODEL & CODE	MODEL YEAR	TEST VEH NO.	ACTUAL	EQUIV TO	DIFFERENT?	H	V	TEST SPD km/h	ACTUAL	ADJ
63033 KIA SPECTRA	2002	C20507	AP1	AP1		130	27	23.6	684	713
			AP2	AP2		203	50	23.9	847	856
			AP3	AP3		159	45	23.9	871	880
			BP1	BP1		270	17	23.7	575	593
			BP4	BP4		140	-2	23.5	585	617
			RP1	RP1		55	33	23.8	601	614
			RP2	RP2		270	1	23.7	450	464
			SR1	SR1		270	20	23.9	767	775
			SR3B	SR3B		270	25	24.0	526	526
			UR4	UR4		90	40	23.9	763	771

EQUIVALENT PRE-STANDARD VEHICLE:

MAKE-MODEL & CODE	MODEL YEAR	TEST VEH NO.	ACTUAL	EQUIV TO	DIFFERENT?	H	V	TEST SPD km/h	ACTUAL	ADJ
63031 KIA SEPHIA	1997	PV0500	AP1	AP1		105	40	24.1	1526	1510
			AP2	AP2		207	46	23.7	960	991
			AP3	AP3		153	44	24.2	1344	1316
			BP1	BP1		270	17	24.1	770	762
			BP4	BP4		140	-6	23.4	829	883
			RP1	RP1		45	39	24.1	988	978
			RP2	RP2		270	0	24.1	682	675
			SR1	SR1		270	42	24.1	970	960
			SR32	SR3B	*	270	50	23.8	604	617
			UR5	UR4	*	90	50	23.8	933	953

NOTE: IN THE ANALYSIS, RESULTS FOR LOCATIONS SR32 AND UR6 ON THE PRE-STANDARD VEHICLE WILL BE COMPARED TO SR3B AND UR4, RESPECTIVELY, ON THE COMPLIANCE TEST.

APPENDIX C

COMPARISON OF HIC SCORES FOR PRE-STANDARD AND POST-STANDARD VEHICLES ON HEAD IMPACT TESTS AT MATCHING TARGET LOCATIONS

Column 1 — Name of the make-model preceded by the 5-digit make-model code used in NHTSA evaluations

Column 2 — "Generalized" target location. This is the actual target on the pre- and post-standard vehicles, if they are the same. Otherwise, it is the actual location on the post-standard vehicle (and the pre-standard vehicle was tested at a nearby, corresponding location); in two tests on the Toyota Camry, it is the actual location on the pre-standard vehicle (and a compliance test was conducted at a nearby, corresponding location).

Column 3 — HIC(d) on the pre-standard vehicle, speed-adjusted to 24 km/h.

Column 4 — HIC(d) on the post-standard vehicle, speed-adjusted to 24 km/h.

Column 5 — HIC(d) improvement from pre-standard to post-standard (positive = improvement, negative = became worse).

Columns 6-7 — Model years of the pre-standard and post-standard (compliance-tested) vehicles, respectively, for this target.

Columns 8-10 — Identifies the actual target locations tested. Column 8 is the actual target on the pre-standard vehicle. Column 9 is the actual target on the compliance test. Column 10 flags with a * when the actual targets are not identical.

Columns 11-12 — Horizontal impact angles on the pre- and post-standard vehicles, respectively. Horizontal impact angle measures degrees away from straight-ahead.

Columns 13-14 — Vertical impact angles on the pre- and post-standard vehicles, respectively. Vertical angle measures degrees away from level (positive = upwards, negative = downwards).

Column 15 — Quality of the match between the pre- and post-standard impacts. If the target locations are identical, and the horizontal impact angles differ by 15 degrees or less and the vertical impact angles differ by 15 degrees or less, it is an "excellent match"; if not, "other match."

43

MAKE-MODEL & CODE	GEN TARGET LOC	HIC(d) (ADJUSTED) PRE	POST	IMPROV	MODEL YEAR PRE	POST	ACTUAL TARGET PRE	POST	DIFFER?	HORIZ ANGLE PRE	POST	VERTICAL ANGLE PRE	POST	MATCH QUALITY
2312 JEEP GRAND CHEROKEE	AP1	1390	748	642	1996	1999	AP1	AP1		105	110	31	29	EXCELLENT MATCH
	AP2	1365	885	480	1996	1999	AP2	AP2		201	202	41	46	EXCELLENT MATCH
	AP3	1452	724	727	1996	1999	AP3	AP3		160	160	39	40	EXCELLENT MATCH
	BP3	683	597	86	1996	1999	BP3	BP3		90	80	-3	-8	EXCELLENT MATCH
	BP4	774	679	95	1996	1999	BP4	BP4		215	270	-6	-9	OTHER MATCH
	OP1	798	509	289	1996	1999	OP1	OP1		270	270	6	8	EXCELLENT MATCH
	OP2	842	505	337	1996	1999	OP2	OP2		90	90	1	-6	EXCELLENT MATCH
	SR1	1442	674	768	1996	1999	SR1	SR1		270	270	35	32	EXCELLENT MATCH
	SR2A	1095	804	291	1996	1999	SR2A	SR2A		270	270	33	33	EXCELLENT MATCH
	UR4	1155	872	283	1996	1999	UR4	UR4		90	90	50	41	EXCELLENT MATCH
7020 DODGE NEON	AP1	827	606	221	1997	2000	AP1	AP1		130	144	42	41	EXCELLENT MATCH
	AP2	987	712	275	1997	2000	AP2	AP2		205	199	44	50	EXCELLENT MATCH
	AP3	1148	902	246	1997	2000	AP3	AP3		154	161	41	48	EXCELLENT MATCH
	BP1	894	658	236	1997	2000	BP1	BP1		90	90	11	22	EXCELLENT MATCH
	BP3	616	578	38	1997	2000	BP3	BP3		280	286	2	-2	EXCELLENT MATCH
	BP4	537	957	-420	1997	2000	BP4	BP4		157	150	-6	-6	EXCELLENT MATCH
	RP2	890	618	272	1997	2000	RP2	RP2		40	70	21	20	OTHER MATCH
	SR1	1144	703	441	1997	2000	SR1	SR1		270	270	33	43	EXCELLENT MATCH
	SR2A	987	809	178	1997	2000	SR2A	SR2A		90	90	27	43	OTHER MATCH
	UR2	1055	720	335	1997	2000	UR2	UR2		270	270	25	28	EXCELLENT MATCH
7406 DODGE GRAND CARAVAN	AP1	1767	573	1194	1996	2002	AP1	AP1		115	140	50	28	OTHER MATCH
	AP2	1432	714	717	1996	2002	AP2	AP2		205	201	46	47	EXCELLENT MATCH
	AP3	1614	515	1098	1996	2002	AP3	AP3		157	160	42	44	EXCELLENT MATCH
	BP2	837	627	210	1996	2002	BP2	BP2		90	90	13	6	EXCELLENT MATCH
	BP3	745	679	66	1996	2002	BP3	BP3		270	270	5	0	EXCELLENT MATCH
	OP2	716	694	22	1996	2002	OP2	OP2		270	90	3	-4	OTHER MATCH
	SD	772	599	173	1996	2002	SD	SD		90	90	50	47	EXCELLENT MATCH
	SR1	1481	675	806	1996	2002	SR1	SR1		270	270	28	32	EXCELLENT MATCH
	UR2	783	470	313	1996	2002	UR2	UR2		270	270	50	45	EXCELLENT MATCH
	UR4	1363	624	739	1996	2002	UR4	UR4		90	90	49	32	OTHER MATCH

MAKE-MODEL & CODE	GEN TARGET LOC	HIC(d) (ADJUSTED)			MODEL YEAR		ACTUAL TARGET		DIFFER?	HORIZ ANGLE		VERTICAL ANGLE		MATCH QUALITY
		PRE	POST	IMPROV	PRE	POST	PRE	POST		PRE	POST	PRE	POST	
12212 FORD F150 SUPERCAB PK	AP1	1220	839	381	1998	2002	AP1	AP1		110	141	31	8	OTHER MATCH
	AP2	1403	589	814	1998	2002	AP2	AP2		201	202	38	30	EXCELLENT MATCH
	AP3	455	708	-254	1998	2002	AP3	AP3		158	150	38	41	EXCELLENT MATCH
	RH	443	657	-214	1998	2002	RH	RH		0	0	40	50	EXCELLENT MATCH
	RP2	706	617	89	1998	2002	RP2	RP2		270	85	0	8	OTHER MATCH
	SR2A	1095	512	583	1998	2002	SR2A	SR2A		270	270	35	32	EXCELLENT MATCH
	SR3	686	884	-198	1998	2002	SR31	SR3	*	270	90	24	35	OTHER MATCH
	UR1	1263	734	529	1998	2002	UR1	UR1		270	270	50	50	EXCELLENT MATCH
	UR2	731	643	88	1998	2002	UR2	UR2		270	270	50	38	EXCELLENT MATCH
	UR6	993	654	339	1998	2002	UR3	UR6	*	270	45	50	50	OTHER MATCH
12302 FORD EXPLORER 4DR	AP1	903	843	59	1998	2002	AP1	AP1		110	130	33	38	OTHER MATCH
	AP2	893	805	87	1998	2002	AP2	AP2		205	219	40	40	EXCELLENT MATCH
	AP3	914	388	526	1998	2002	AP3	AP3		155	134	35	24	OTHER MATCH
	BP1	997	738	259	1998	2002	BP1	BP1		90	90	11	6	EXCELLENT MATCH
	BP2	669	867	-198	1998	2002	BP2	BP2		270	270	4	9	EXCELLENT MATCH
	BP4	537	545	-9	1998	2002	BP4	BP4		140	143	-5	-10	EXCELLENT MATCH
	FH2	426	482	-56	1998	2002	FH2	FH2		180	90	50	50	OTHER MATCH
	OP1	583	492	91	1998	2002	OP1	OP1		90	90	1	5	EXCELLENT MATCH
	SR1	1224	813	411	1998	2002	SR1	SR1		270	270	31	39	EXCELLENT MATCH
	UR1	1142	443	699	1998	2002	UR1	UR1		270	270	50	50	EXCELLENT MATCH
	UR2	958	476	481	1998	2002	UR2	UR2		270	360	50	50	OTHER MATCH

MAKE-MODEL & CODE	GEN TARGET LOC	HIC(d) (ADJUSTED)			MODEL YEAR		ACTUAL TARGET		DIFFER?	HORIZ ANGLE		VERTICAL ANGLE		MATCH QUALITY
		PRE	POST	IMPROV	PRE	POST	PRE	POST		PRE	POST	PRE	POST	
12402 FORD WINDSTAR WAGON	AP1	1015	711	304	1998	1999	AP1	AP1		155	147	43	25	OTHER MATCH
	AP2	1450	487	963	1998	1999	AP2	AP2		205	213	46	42	EXCELLENT MATCH
	AP3	1702	652	1050	1998	1999	AP3	AP3		155	144	40	35	EXCELLENT MATCH
	BP1	1014	489	525	1998	1999	BP1	BP1		90	270	21	22	OTHER MATCH
	BP2	746	766	-20	1998	1999	BP2	BP2		90	90	0	5	EXCELLENT MATCH
	BP4	600	713	-113	1998	1999	BP4	BP4		207	270	-8	70	OTHER MATCH
	FH1	655	652	3	1998	1999	FH1	FH1		180	180	50	50	EXCELLENT MATCH
	OP1	697	637	60	1998	1999	OP1	OP1		270	90	0	-3	OTHER MATCH
	OP2	603	684	-81	1998	1999	OP2	OP2		90	270	-1	5	OTHER MATCH
	RP1	780	715	65	1998	1999	RP1	RP1		40	317	44	50	OTHER MATCH
	RP2	690	870	-179	1998	1999	RP2	RP2		345	90	13	45	OTHER MATCH
	SD	825	698	127	1998	1999	SD	SD		90	270	28	30	OTHER MATCH
	SR1	884	898	-14	1998	1999	SR1	SR1		270	270	38	40	EXCELLENT MATCH
	SR2A	792	870	-78	1998	1999	SR2A	SR2A		270	90	35	45	OTHER MATCH
	SR3	734	764	-30	1998	1999	SR31	SR3	*	270	90	39	50	OTHER MATCH
	UR	999	503	495	1998	1999	UR2	UR	*	270	90	41	38	OTHER MATCH
18002 BUICK LeSABRE	AP1	1075	674	401	1998	2001	AP1	AP1		250	250	38	30	EXCELLENT MATCH
	AP3	1352	710	641	1998	2001	AP3	AP3		149	139	43	39	EXCELLENT MATCH
	BP1	604	693	-89	1998	2001	BP1	BP1		90	90	5	20	EXCELLENT MATCH
	BP3	575	820	-245	1998	2001	BP3	BP3		275	281	3	-6	EXCELLENT MATCH
	BP4	547	654	-107	1998	2001	BP4	BP4		161	125	-4	-5	OTHER MATCH
	RP1	722	849	-128	1998	2001	RP1	RP1		290	285	30	29	EXCELLENT MATCH
	RP2	581	708	-127	1998	2001	RP2	RP2		67	46	23	26	OTHER MATCH
	SR2B	1073	769	304	1998	2001	SR2B	SR2B		90	90	45	50	EXCELLENT MATCH
	SR3	750	733	17	1998	2001	SR31	SR3	*	270	270	31	50	OTHER MATCH
	UR	897	663	234	1998	2001	UR2	UR	*	270	270	37	50	OTHER MATCH
20212 CHEVY C-1500/ SILVERADO X-CAB PK	AP1	1308	666	641	1998	1999	AP1	AP1		248	249	35	25	EXCELLENT MATCH
	AP2	1216	462	754	1998	1999	AP2	AP2		161	119	29	35	OTHER MATCH
	AP3	1057	551	505	1998	1999	AP3	AP3		201	216	32	32	EXCELLENT MATCH
	RP1	883	742	141	1998	1999	RP1	RP1		270	274	7	22	EXCELLENT MATCH
	RP2	581	669	-87	1998	1999	RP2	RP2		90	90	5	10	EXCELLENT MATCH
	SR2A	1061	699	362	1998	1999	SR2A	SR2A		270	270	23	48	OTHER MATCH
	UR	543	694	-150	1998	1999	UR2	UR	*	270	270	50	50	OTHER MATCH

46

MAKE-MODEL & CODE	GEN TARGET LOC	HIC(d) (ADJUSTED) PRE	POST	IMPROV	MODEL YEAR PRE	POST	ACTUAL TARGET PRE	POST	DIFFER?	HORIZ ANGLE PRE	POST	VERTICAL ANGLE PRE	POST	MATCH QUALITY
20302 CHEVY BLAZER / TRAILBLAZER 4DR	AP1	1206	602	604	1996	2002	AP1	AP1		113	246	25	38	OTHER MATCH
	AP2	1307	785	522	1996	2002	AP2	AP2		206	143	38	40	OTHER MATCH
	AP3	1466	986	481	1996	2002	AP3	AP3		155	214	36	36	OTHER MATCH
	BP1	683	728	-45	1996	2002	BP1	BP1		90	90	27	21	EXCELLENT MATCH
	BP3	794	886	-93	1996	2002	BP3	BP3		270	281	-2	-6	EXCELLENT MATCH
	BP4	836	539	297	1996	2002	BP4	BP4		110	122	-10	-10	EXCELLENT MATCH
	OP1	753	498	255	1996	2002	OP1	OP1		90	90	5	-10	EXCELLENT MATCH
	SR2A	1250	713	537	1996	2002	SR2A	SR2A		270	270	43	50	EXCELLENT MATCH
	SR3	665	545	120	1996	2002	SR33	SR3	*	90	90	30	50	OTHER MATCH
	UR	727	593	133	1996	2002	UR2	UR	*	270	270	50	50	OTHER MATCH
35043 NISSAN SENTRA	AP1	926	874	52	1998	2001	AP1	AP1		113	125	21	32	EXCELLENT MATCH
	AP2	1311	615	696	1998	2001	AP2	AP2		203	201	47	48	EXCELLENT MATCH
	AP3	1584	558	1026	1998	2001	AP3	AP3		157	159	44	44	EXCELLENT MATCH
	BP1	758	763	-5	1998	2001	BP1	BP1		90	90	12	22	EXCELLENT MATCH
	BP2	643	685	-42	1998	2001	BP2	BP2		90	90	12	11	EXCELLENT MATCH
	BP3	778	591	187	1998	2001	BP3	BP3		284	279	-4	-3	EXCELLENT MATCH
	RP1	659	576	83	1998	2001	RP1	RP1		90	90	14	15	EXCELLENT MATCH
	SR1	884	757	127	1998	2001	SR1	SR1		270	270	20	46	OTHER MATCH
	SR2A	689	830	-141	1998	2001	SR2A	SR2A		270	270	18	37	OTHER MATCH
	UR	955	832	123	1998	2001	UR2	UR	*	270	270	38	36	OTHER MATCH
37032 HONDA ACCORD	AP2	1047	580	467	1996	1999	AP2	AP2		206	246	50	40	OTHER MATCH
	BP2	830	652	178	1996	AVG 99 & 03	BP2	BP2		90	90	20	16	EXCELLENT MATCH
	RP2	600	662	-62	1996	AVG 99 & 03	RP2	RP2		80	70	8	22	EXCELLENT MATCH
	SR2	1248	532	716	1996	1999	SR2A	SR2	*	270	90	29	50	OTHER MATCH
	UR2	757	765	-8	1996	2003	UR2	UR2		270	90	38	36	OTHER MATCH
	UR3	905	505	399	1996	2003	UR3	UR3		270	90	50	50	OTHER MATCH
	UR4	841	622	219	1996	2003	UR4	UR4		90	90	46	43	EXCELLENT MATCH
	UR5	825	630	195	1996	2003	UR5	UR5		90	180	39	48	OTHER MATCH

MAKE-MODEL & CODE	GEN TARGET LOC	HIC(d) (ADJUSTED) PRE	POST	IMPROV	MODEL YEAR PRE	POST	ACTUAL TARGET PRE	POST	DIFFER?	HORIZ ANGLE PRE	POST	VERTICAL ANGLE PRE	POST	MATCH QUALITY
49032 TOYOTA COROLLA	AP1	623	703	-80	1998	2003	AP1	AP1		109	124	21	17	EXCELLENT MATCH
	AP3	647	578	69	1998	2003	AP3	AP3		157	155	42	50	EXCELLENT MATCH
	BP1	695	637	58	1998	2003	BP1	BP1		90	90	10	12	EXCELLENT MATCH
	BP2	589	585	4	1998	2003	BP2	BP2		270	270	8	4	EXCELLENT MATCH
	BP3	610	512	98	1998	2003	BP3	BP3		83	88	-1	-8	EXCELLENT MATCH
	RP1	663	553	110	1998	2003	RP1	RP1		82	60	18	16	OTHER MATCH
	RP2	772	578	194	1998	2003	RP2	RP2		280	296	13	17	OTHER MATCH
	SR1	836	582	254	1998	2003	SR1	SR1		270	270	27	22	EXCELLENT MATCH
	SR2A	632	390	242	1998	2003	SR2A	SR2A		90	90	17	10	EXCELLENT MATCH
	UR	816	760	56	1998	2003	UR2	UR	*	270	270	37	46	OTHER MATCH
49040 TOYOTA CAMRY	AP1	753	754	-2	1998	AVG 99 & 02	AP1	AP1		108	113	32	25	EXCELLENT MATCH
	AP2	936	707	229	1998	AVG 99 & 02	AP2	AP2		204	221	47	47	OTHER MATCH
	AP3	990	647	343	1998	AVG 99 & 02	AP3	AP3		157	133	43	43	OTHER MATCH
	BP1	951	633	318	1998	AVG 99 & 02	BP1	BP1		270	270	24	14	EXCELLENT MATCH
	BP2	922	742	180	1998	2002	BP2	BP2		270	270	11	11	EXCELLENT MATCH
	BP4	728	716	12	1998	AVG 99 & 02	BP4	BP4		160	114	-8	-4	OTHER MATCH
	RP1	640	612	28	1998	AVG 99 & 02	RP1	RP1		270	291	22	23	OTHER MATCH
	RP2	730	840	-110	1998	AVG 99 & 02	RP2	RP2		70	73	21	32	EXCELLENT MATCH
	SR2A	880	881	-1	1998	1999	SR2A	SR2A		270	270	28	30	EXCELLENT MATCH
	SR31	625	509	116	1998	2002	SR31	SR3	*	270	270	34	35	OTHER MATCH
	SR32	664	641	23	1998	1999	SR32	SR3	*	90	90	38	32	OTHER MATCH
	UR	1012	863	149	1998	2002	UR5	UR	*	90	90	39	45	OTHER MATCH
49202 TOYOTA TACOMA XTRACAB	AP1	1121	595	525	1997	2003	AP1	AP1		137	150	35	20	EXCELLENT MATCH
	AP2	931	545	386	1997	2003	AP2	AP2		217	230	30	30	EXCELLENT MATCH
	AP3	707	373	334	1997	2003	AP3	AP3		144	130	30	30	EXCELLENT MATCH
	FH1	559	682	-123	1997	2003	FH1	FH1		180	180	50	50	EXCELLENT MATCH
	RH	629	414	215	1997	2003	RH	RH		0	0	31	34	EXCELLENT MATCH
	RP1	1059	627	431	1997	2003	RP1	RP1		45	52	29	17	EXCELLENT MATCH
	RP2	1076	715	361	1997	2003	RP2	RP2		270	78	5	1	OTHER MATCH
	SR1	1086	427	659	1997	2003	SR1	SR1		270	270	33	25	EXCELLENT MATCH
	SR2	805	512	293	1997	2003	SR2A	SR2	*	270	90	31	31	OTHER MATCH
	UR	1029	731	298	1997	2003	UR5	UR	*	90	276	50	44	OTHER MATCH

MAKE-MODEL & CODE	GEN TARGET LOC	HIC(d) (ADJUSTED)			MODEL YEAR		ACTUAL TARGET		DIFFER?	HORIZ ANGLE		VERTICAL ANGLE		MATCH QUALITY
		PRE	POST	IMPROV	PRE	POST	PRE	POST		PRE	POST	PRE	POST	
63031 KIA SEPHIA/	AP1	1510	713	797	1997	2002	AP1	AP1		105	130	40	27	OTHER MATCH
63033 KIA SPECTRA	AP2	991	856	135	1997	2002	AP2	AP2		207	203	46	50	EXCELLENT MATCH
	AP3	1316	880	436	1997	2002	AP3	AP3		153	159	44	45	EXCELLENT MATCH
	BP1	762	593	169	1997	2002	BP1	BP1		270	270	17	17	EXCELLENT MATCH
	BP4	883	617	267	1997	2002	BP4	BP4		140	140	-6	-2	EXCELLENT MATCH
	RP1	978	614	364	1997	2002	RP1	RP1		45	55	39	33	EXCELLENT MATCH
	RP2	675	464	211	1997	2002	RP2	RP2		270	270	0	1	EXCELLENT MATCH
	SR1	960	775	185	1997	2002	SR1	SR1		270	270	42	20	OTHER MATCH
	SR3B	617	526	91	1997	2002	SR32	SR3B	*	270	270	50	25	OTHER MATCH
	UR4	953	771	182	1997	2002	UR5	UR4	*	90	90	50	40	OTHER MATCH